T0140815

Defect engineering in transition metal based nitride thin films by energetic treatment during deposition

Von der Fakultät für Werkstoffwissenschaft und Werkstofftechnologie

der Technischen Universität Bergakademie Freiberg

genehmigte

DISSERTATION

zur Erlangung des akademischen Grades

Doktor-Ingenieur

Dr.-Ing.

vorgelegt

von Dipl.-Ing. Christina Wüstefeld

geboren am 11.09.1981 in Meißen.

Gutachter: Prof. Dr. rer. nat. habil. David Rafaja, Freiberg

Prof. RNDr. Radomír Kužel, CSc., Prag

Tag der Verleihung: 19.11.2015

Bibliografische Information der Deutschen Nationalbibliothek

Die Deutsche Nationalbibliothek verzeichnet diese Publikation in der
Deutschen Nationalbibliografie; detaillierte bibliografische Daten sind
im Internet über http://dnb.d-nb.de abrufbar.

ISBN 978-3-8325-4267-2

Logos Verlag Berlin GmbH
Comeniushof, Gubener Str. 47,
10243 Berlin
Tel.: +49 (0)30 42 85 10 90
Fax: +49 (0)30 42 85 10 92
INTERNET: http://www.logos-verlag.de

Danksagung

Diese Arbeit entstand während meiner Tätigkeit als wissenschaftliche Mitarbeiterin im Forschungsprojekt der sächsischen Landesexzellenzinitiative auf der Basis von Spitzentechnologieclustern: „Funktionales Strukturdesign neuer Hochleistungswerkstoffe durch Atomares Design und Defekt-Engineering (ADDE)" sowie im DFG-Projekt „Ausbildung und thermische Stabilität von nanoskaligen Domänen in ausgewählten ternären und quarternären dünnen Schichten" am Institut für Werkstoffwissenschaft der Technischen Universität Bergakademie Freiberg.

Ohne die Unterstützung vieler Menschen hätte es diese Arbeit nicht gegeben. Ich möchte mich bei allen bedanken, die mich bei der Anfertigung der Arbeit auf unterschiedliche Art und Weise unterstützt haben.

Mein größter Dank gilt Prof. Dr. David Rafaja für die wissenschaftliche Betreuung der Arbeit, die wertvollen Diskussionen und Anregungen sowie für seine Motivation und Geduld. Seine kontinuierliche Unterstützung in den vergangenen Jahren hat entscheidend zu dieser Arbeit beigetragen.

Ich danke außerdem Prof. Dr. Radomír Kužel für die Übernahme des Zweitgutachtens.

Ein sehr großer Dank geht an Dr. Mykhaylo Motylenko für seine zahlreichen sorgfältigen TEM-Untersuchungen, die vielen hilfreichen Diskussionen und für seinen Enthusiasmus. Diese Zusammenarbeit war sehr bereichernd und wunderbar!

Ein besonderer Dank geht an meinen Bürokollegen Dr. Milan Dopita für seine fachliche und persönliche Unterstützung. Seine positive Einstellung und sein Humor trugen zu einer sehr guten Arbeitsatmosphäre bei.

Prof. Dr. Arutiun Ehiasarian vom Nanotechnology Centre for PVD Research der Sheffield Hallam Universität danke ich für die Kooperation, die Ermöglichung eines Forschungsaufenthaltes in seiner Arbeitsgruppe und für die Unterstützung bei der Abscheidung der Cr-Al-Si-N- Schichten.

Dr. Martin Kathrein und Dr. Christoph Czettl von Ceratizit Austria GmbH und Claude Michotte von Ceratizit Luxembourg S.à.r.l. danke ich für die Abscheidung und Bereitstellung der Schichten basierend auf Ti-Al-N.

Dr. Carsten Baehtz danke ich für seine Hilfe bei der Durchführung der *in situ* Synchrotron-Experimente an der Beamline ROBL BM20 an der Europäischen Synchrotron-Strahlungsquelle (ESRF) in Grenoble.

Weiterhin danke ich allen meinen Kollegen vom Institut für Werkstoffwissenschaft für die Unterstützung in diversen experimentellen, präparativen und technischen Angelegenheiten. Besonders danke ich sehr: Gerhard Schreiber, Dr. Dietrich Heger, Dr. Volker Klemm, Astrid Leuteritz, Roswitha Popp, Dr. Mario Kriegel, Florian Hanzig, Ulrike Ratayski, Uwe Gubsch, Nora Leonhardt, Brigitte Bleiber, Angelika Müller, Katrin Becker und Uwe Schönherr.

Ganz herzlich danke ich Andreas und meiner Familie für ihre vielfältige, unendliche Unterstützung, Ermutigung und ihren Rück(en)halt!

Contents

List of Abbreviations

ad	As-deposited
AFM	Antiferromagnetic
BF	Bright field
CAE	Cathodic arc evaporation
CCD	Charge-coupled device
dc	Direct current
DF	Dark field
EDS	Energy dispersive spectroscopy
EEL	Electron energy loss
EELS	Electron energy loss spectroscopy
ELNES	Electron loss near edge spectroscopy
EPMA / WDS	Electron probe microanalysis with wavelength-dispersive X-ray spectroscopy
Fcc	Face centred cubic
FFT	Fast Fourier transformation
FIB	Focused ion beam
GAXRD	Glancing angle X-ray diffraction
GDOES	Glow discharge optical emission spectroscopy
HAADF	High-angle annular dark field
HIPIMS	High power impulse magnetron sputtering
HRTEM	High resolution transmission electron microscopy
HRSTEM	High resolution scanning transmission electron microscopy
HRXRD	High resolution X-ray diffraction
HT-GAXRD	High temperature glancing angle X-ray diffraction
MFP	Mean free path of inelastic scattering
MS	Magnetron sputtering
OR	Orientation relationship
PM	Powder metallurgical
SAED	Selected area electron diffraction
STEM	Scanning transmission electron microscopy
TEM	Transmission electron microscopy
UBM	Unbalanced magnetron sputtering

XEC	X-ray elastic constant
XRD	X-ray diffraction
w	Wurtzitic
ZL	Zero loss in EEL spectra

List of Symbols

a, c	Lattice parameters
a_0	Stress-free lattice parameter
$a_\perp$	Lattice parameter perpendicular to the film plane
$a_\parallel$	Lattice parameter parallel to the film plane
a^{hkl}	Lattice parameter of the lattice plane with the Miller Indices (hkl)
a_ψ^{hkl}	Lattice parameter of the lattice plane with the Miller Indices (hkl) measured at a certain angle ψ
$\vec{a}^*, \vec{b}^*, \vec{c}^*$	Unit-cell translation vectors in reciprocal space
a_i	Parameters of the analytical scattering factor functions
A	Anisotropy factor
A_c	Contact area in an indentation experiment
A_{aw}	Atomic weight
A_1, A_2, B_1, B_2	Coefficients of the function $a_\psi^{hkl}(\Gamma, \psi)$
A_p	Projected contact area
b_{ebeam}	Broadening of the diameter of the electron beam in a TEM foil
b_i	Parameters of the analytical scattering factor functions
$\vec{B}$	Magnetic field vector
$\vec{B}_{ebeam}$	Direction of the electron beam in TEM
c_0	Parameter of the analytical scattering factor function
c_A, c_B	Atomic concentration of the element A and of the element B
C_i	Constants describing the real indenter tip
C_{ijkl}	Stiffness tensor (4th rank)
C_{ij}, C_{kl}	Stiffness tensor in matrix notation (2nd rank)
C_{11}, C_{12}, C_{44}	Stiffnesses of a single crystal
C_s	Spherical aberration
d	Interplanar spacing
d_0	Unstrained interplanar spacing
d^{hkl}	Interplanar spacing of a plane with diffraction indices hkl
d_0^{hkl}	Unstrained interplanar spacing of a plane with diffraction indices hkl

$d_{\phi\psi}^{hkl}$	Interplanar spacing of a plane with diffraction indices hkl measured at a certain rotation angle ϕ and tilt angle ψ
d_i	Diameter of the inner circle of the crater from ball crater test
d_L	Coating thickness
d_{ENTA}	Diameter of the spectrometer entrance aperture in STEM
D_o	Diameter of the outer circle of the crater from ball crater test
$\Delta \, pix_i$	Difference in the CCD camera pixel number
$e^{\ominus}$	Elementary charge of an electron
e	Microstrain
E	Young's modulus
E_0	Energy of the incident electrons in TEM
E_m	Mean energy loss in EELS
E_{IT}	Indentation modulus
E_i	Modulus of the indenter
E_{iES}	Increment of the energy shift in EELS
E_r	Reduced elastic modulus
E_{i0}	Kinetic energy of a cathodic arc ion
$\Delta E_{i,kin}$	Gain of kinetic energy of a cathodic arc ion
E_{ic}	Kinetic energy of a cathodic arc ion by image charge acceleration
E_c	Cohesive energy
E_{exc}	Excitation energy of bound electrons
E_{ion}	Cumulative ionization energy
E_{total}	Total energy brought by an ion to the substrate
f	Misfit
f_j	Atomic scattering factor
f_{Me}	Atomic scattering factor of a metal
f_{Al}	Atomic scattering factor of Al
f_{Ti}	Atomic scattering factor of Ti
f_N	Atomic scattering factor of N
f_{EELS}	Calibration factor for the EELS spectra
F	Relativistic correction factor
F_{hkl}	Structure factor
F_{max}	Maximum load

$\vec{g}$	Vector of the reciprocal lattice
h, k, l	Miller Indices
h_0	Depth of the residual hardness impression
h_c	Contact depth during nanoindentation
h_{max}	Maximum penetration depth during hardness measurement
h_s	Surface displacement during nanoindentation
h_{sa}	Distance between TEM foil and spectrometer aperture in STEM
H_{IT}	Indentation hardness
$I(hkl)$	Integral intensity of a diffraction line with diffraction indices hkl
I_A	Integral intensity of a characteristic X-ray peak of the element A in EDS
I_B	Integral intensity of a characteristic X-ray peak of the element B in EDS
I_0	Area under the zero-loss peak in an EEL spectrum
I_t	Total area under the whole EEL spectrum
I_{TiN}^N	Integral intensity of the N K edge within an integration width of 30 eV measured for fcc-TiN
$\vec{j}$	Current density
k_{AB}	Cliff Lorimer factor
m_{WH}	Slope in the Williamson Hall plot
m_i	Mass of an ion
P_1, P_2, P_3, P_4	Fitting parameters
P_{Al-Si}	Power applied to the Al-Si target
P_{Cr}	Power applied to the Cr target
$P(Cr)$	Power ratio applied to the Cr target
q	Modulus of the diffraction vector
$\vec{q}$	Diffraction vector in XRD
Q	Charge state number of an ion
S	Contact stiffness
s_1^{hkl}, s_2^{hkl}	X-ray elastic constants
S_{ijkl}	Compliance tensor (4th rank)
S_{ij}, S_{kl}	Compliance tensor in matrix notation (2nd rank)
S_{11}, S_{12}, S_{44}	Compliance constants of a single crystal
S_1, S_2	Elastic constants for elastically isotropic crystallites
t	Thickness of the TEM foil
T_S	Substrate temperature

$T_{measure}$	Measuring temperature
U_B	Substrate bias voltage
x	Al concentration
x_e	Penetration depth of X-rays
ΔV_{sheath}	Potential difference between the substrate and the plasma potential
Z	Atomic number
γ	Angle of incidence for GAXRD
Γ	Orientation factor
ε	Macroscopic lattice strain
$\varepsilon_{\phi\psi}$	Lattice strain depending on the angles ϕ and ψ
ε_{ij}	Strain tensor
ϵ	Geometric constant of the indenter
θ	Half of the diffraction angle 2θ
θ^{hkl}	Half of the diffraction angle 2θ of a certain hkl plane
λ	Wavelength
λ_{MFP}	Mean free path of inelastic scattering
μ	Linear absorption coefficient
ν	Poisson's ratio
ν_i	Poisson's ratio of the indenter
ρ	Density
σ	Residual stress
σ_g	Residual stress contribution from the layer growing process
σ_{kl}	Stress tensor
σ_{th}	Thermal stresses
σ_p	Stresses caused by phase transformation
v_{io}	Velocity of an ion
φ	Angle between two lattice planes
ϕ	Angle describing the rotation around the surface normal
ϕ_S	Work function of the substrate side
ψ	Angle between the diffraction vector and the sample surface normal
ω	Angle between the primary beam and the sample surface

1 Introduction

Transition metal (TM) nitride thin films on the basis of TM-Al-N are important for many industrial applications as wear protective coatings due to their unique properties like high hardness, thermal stability, good oxidation resistance and abrasion resistance. Prominent examples of industrially used ternary transition metal nitride coatings are Ti-Al-N and Cr-Al-N films. Coatings based on Ti-Al-N are used mainly in metal cutting applications offering high cutting speeds and increased tool life due to their thermal stability [1]. Coatings based on Cr-Al-N offer a good wear and oxidation resistance as well as resistance against electrochemical corrosion. Thus, these coatings are used beside machining applications [2] also for molding and forming dies [3, 4].

Ti-Al-N and Cr-Al-N films are known as hard coatings since the middle of the 1980s [5-7] and the beginning of 1990s [8, 9], respectively. Still, several attempts are made to improve the coatings based on both systems in order to increase the lifetime of the coated tools and the machining speed. The concepts for the improvement of the coating properties comprise the addition of further alloying and doping elements, the growth of multilayers and the adjustment of the deposition parameters. These approaches are used to modify the elemental composition and the microstructure especially the phase composition, crystallite size, macroscopic stress, interfaces, types and density of defects in order to obtain certain coating properties.

The aim of this study was the investigation of the influence of the energetic treatment during the deposition of transition metal based nitride coatings on their microstructure and properties. The research was performed on industrially relevant Ti-Al-N monolithic coatings, Ti-Al-N / Al-Ti-Ru-N multilayer coatings and Cr-Al-Si-N monolithic coatings.

Within this study, the energetic treatment was realized by ion bombardment of the growing film with film forming ions during physical vapour deposition (PVD) of the coatings. The energy of the ions arriving at the substrate is determined by the difference of the plasma and substrate surface potential, which can be controlled by the application of a substrate bias voltage. In that way, the bias voltage is an effective parameter to manipulate the energetic treatment of the growing film. Beside the control of the acceleration of the ions towards the substrate using different bias voltages, the fraction of the film forming ions, which is provided by the deposition method, influences the energetic treatment of the growing film. Thus, three PVD deposition methods, namely magnetron sputtering (MS), high power impuls magnetron sputtering (HIPIMS) and cathodic arc evaporation (CAE), were used within this work. The CAE process is characterized by the highest amount of film forming ions whereas magnetron sputter deposition provides the lowest amount of film forming ions.

The main part of this study is focused on the microstructure characterization of industrially relevant Ti-Al-N monolayer coatings with different Al content and Ti-Al-N / Al-Ti-(Ru)-N multilayer coatings with different Ru addition which were deposited by CAE with different bias voltages. The influence of the energetic treatment during deposition, as controlled by the applied substrate bias voltage, as well as the influence of the chemical composition on the microstructure and properties of both coating types was investigated in the as-deposited state and in the course of a thermal treatment up to 950°C. The microstructural characterization of the industrially relevant coatings based on Ti-Al-N are complemented by the analysis of the heteroepitaxial growth of a w-AlN layer which was deposited onto two differently oriented fcc-TiN seed layers.

In a second part of this work, the influence of a different flux of ionized film forming species on the microstructure and properties was investigated on the example of Cr-Al-Si-N coatings which were deposited by unbalanced magnetron sputtering (UBM) and HIPIMS. The results from this characterization are compared with the microstructure and properties of Cr-Al-Si-N coatings deposited by CAE which were analyzed previously by Rafaja *et al.* [10].

The microstructure of the coatings was analysed using X-ray diffraction which yielded information about the phase composition, residual stress, elastic anisotropy, lattice parameter, preferred orientation of crystallites, cluster and crystallite size. Transmission electron microscopy was used to study the local orientation relationship, to visualize the cluster size and the bilayer period. Analytical TEM methods, which comprise energy dispersive X-ray spectroscopy and electron energy loss spectroscopy, yielded information about the chemical and phase composition on the nanometre scale. Electron probe microanalysis using wavelength dispersive X-ray spectroscopy and glow discharge optical emission spectroscopy were applied to determine the global chemical composition of the coatings. The microstructure of the films is related to the coatings' hardness which was measured using nanoindentation.

This thesis is organized in the following way. At first the characteristics of the three deposition methods, magnetron sputtering, HIPIMS and CAE, are described. The relevant coating systems Ti-Al-N, Ti-Al-Ru-N and Cr-Al-Si-N are introduced in the Chapters 3 to 5 and a short literature review is given. The experimental characterization techniques, which were used within this work, are described in Chapter 6. The results and discussion part, shown in Chapter 7, is subdivided according to the three coating types, namely Ti-Al-N monolithic films, Ti-Al-N / Al-Ti-Ru-N multilayers and Cr-Al-Si-N monolithic films. Additionally, the results of the heteroepitaxial growth of w-AlN on fcc-TiN are presented. Finally, the conclusions are given in Chapter 8.

2 Physical vapour deposition technology

The coatings investigated within this thesis were prepared by physical vapour deposition (PVD) using three different methods namely magnetron sputtering (MS), high-power impulse magnetron sputtering (HIPIMS) and cathodic arc evaporation (CAE). These three deposition technologies differ in the flux of the ionized coating species to the growing film. The lowest flux of ions is present during MS and the CAE process is characterized by highest flux of ions. The HIPIMS process ranges in between. This will be presented briefly in the next Sections.

All depositions were done in reactive mode. For the deposition of nitrides, metallic targets were used and nitrogen was provided as gas. The three used PVD techniques are explained briefly in the following Sections 2.1 to 2.3.

2.1 Magnetron sputtering

Magnetron sputtering (MS) is a nonthermal vaporization process at the presence of a plasma. During MS the target, which represents the cathode, is subjected to the bombardment with energetic ions of an inert gas - the so called working gas. Usually argon is used as working gas, because its mass is high enough to ensure adequate sputtering yields and it is less expensive than krypton or xenon. The ions of the working gas are produced in a glow discharge by the inelastic collision of gas atoms with energetic electrons as represented by Eq. (2.1). The ionization energy of argon is 15.7 eV and the maximum ionization probability occurs at an electron energy of about 100 eV [11].

$$e^-(100eV) \; + \; Ar \; \xrightarrow{\; inelastic \;} e^-(< 100eV) \; + \; Ar^+ \; + \; e^- \qquad \textbf{(2.1)}$$

The ions are accelerated to the target because of its negative potential with respect to the plasma. This potential drop determines the energy of the ion at low pressures. At higher pressures, the ions experience physical collisions and charge exchange collisions which yields in a spectrum of energies of ions and neutrals that bombard the target. Due to the momentum transfer from the energetic bombarding ions, atoms of the target surface are ejected that follow the Thompson energy distribution. The Thompson energy distribution is characterized by a maximum at several eV and an energy tail extending to tens and hundreds of eV, which depends on the energy of the bombarding ion. Thus, the energy of most of the atoms arriving at the substrate is in the range of just a few eV and can reach the ejection energy depending whether most of the ejected particles are thermalized by collisions at high pressure (> 0.7 Pa) or not thermalized at low pressure (< 0.7 Pa) [11]. In dc MS only a few sputtered atoms are ionized in the plasma, see e.g. References [12, 13]. As a result of the ion bombardment of the target also secondary electrons are emitted from the target surface. The electrons are responsible for the sustainment of the discharge. By the establishment of strong magnetic fields above the target surface the ejected secondary electrons are trapped close to the target surface where they move spiral-like in the combined electric and magnetic fields. This leads to an increased ionization of Ar atoms by electron atom collision (see Eq. (2.1)) and a dense plasma in the target region. Due to the electric field the positive ions are accelerated to the cathode surface where sputtering takes place. In that way an increased ion bombardment of the target and higher sputtering rates are achieved. Fig. 2.1 illustrates the arrangement of the magnets behind the targets where one pole is positioned at the central axis and the other pole is placed annular at the outer edge of the target. In case of a balanced magnetron (see Fig. 2.1a) the central and outer pole have the same strength so that the magnetic field lines are closed between the two poles. As a result of this, the plasma is confined close to the target region and the ion current density at the substrate is less than 1 mA/cm^2 [14]. Within this work an unbalanced magnetron of type 2 [15] was used. In that case the outer ring of magnets is strengthened relative to the central pole so that not all field lines are closed between the poles. Some field lines are aligned towards the substrate. The secondary electrons can follow these field lines and thus the plasma region expands towards the substrate. As a consequence the substrate ion current densities are increased to 2-10 mA/cm^2 [14]. If a negative substrate bias voltage is applied, Ar ions can be accelerated towards the substrate.

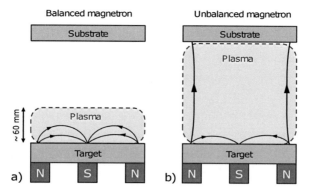

Fig. 2.1: Scheme of the plasma confinement using a conventional balanced magnetron (a) and an unbalanced magnetron (b) (reproduced after Ref. [14]).

2.2   High power impulse magnetron sputtering

High power impulse magnetron sputtering (HIPIMS) is a sputtering ionized physical vapour deposition technology. The HIPIMS plasma is generated by a glow discharge. In contrast to MS, where ions of the sputtered material are rare and mainly ions of the inert sputtering gas are produced, in HIPIMS a significant amount of the sputtered metal species is ionized [16]. During HIPIMS a sharp peak of voltage is applied to the target for a low duration in the range of 50 to 500 μs and a certain frequency ranging from one pulse up to 500 Hz [17]. After some μs after the initiation of the voltage pulse the discharge current rises to a peak value due to an increased electron density. The sputtered atoms cannot penetrate the dense highly energetic electron cloud formed in the sheath without collisions. Hence, most metal ions are formed due to electron impact ionization [12]. Thus, electrons with an energy that corresponds at least to the ionization energy of atoms are required. The ionization is effective, if the electron energy is in the range where the ionization cross section of the metal atoms is large (50 to 100 eV for most atoms), [18]. Additionally, a minor part of metal ions is formed by Penning ionization due to the collision with metastable excited Ar atoms (Ar^*) and charge exchange between metal atoms (M) and Ar ions as illustrated by Eq. (2.2) and Eq. (2.3), respectively:

$$Ar^* + M \longrightarrow M^+ + Ar + e^- \tag{2.2}$$

$$M + Ar^+ \longrightarrow M^+ + Ar \tag{2.3}$$

The ionized metal can escape from the dense plasma region and reach the substrate at long distances [12]. As the sputtered material is ionized close to the target, some metal ions are attracted back to the target surface due to the cathode potential. These ions act as a sputtering particle giving rise to self-sputtering which can run away to very high levels [18]. This effect is considered as one reason for the lower deposition rate of HIPIMS [17]. The reported ionized flux fraction of target atoms using HIPIMS deposition varies from 10 % to > 80 % as stated by Helmersson in his review about ionized PVD [17] which is attributed to different magnetron configurations and applied power densities used by different research groups. Using HIPIMS mostly single charged ions and a significant fraction of ions with 2^+ charge states are obtained [18].

2.3   Cathodic arc evaporation

The following Section about the PVD method **cathodic arc evaporation** (CAE) is mainly based on Anders' book about "Cathodic Arcs: From Fractal Spots to Energetic Condensation" [19]. If the text refers to other references, then the citations are indicated.

In general arc evaporation is characterized by high current-density, low voltage electric current that passes between slightly separated electrodes through a gas or vapor of the electrode material [11]. The potential distribution between the anode and the cathode of a cathodic arc is schematically presented in Fig. 2.2. In Fig. 2.2 the negative potential is plotted to account for the negative charge of the electron [20]. During CAE the target represents the cathode, which is generally a water-cooled solid that is globally considered as a "cold cathode". The electrons of the cathode need to be given the work function energy to free them from the cathode. Emitted electrons are accelerated in the cathode fall. The electrons move to the anode where potential energy is released due to the work function. The arrival rate of the electrons is adjusted by the anode potential drop, which can be positive or negative. In Fig. 2.2 the anode fall is shown as a small barrier for the electrons due to an oversupply of electrons.

During CAE the electron emission and plasma production takes places in so-called non-stationary cathode spots on the target surface. Such a cathode spot is described by Anders [19] as an assembly of emission centers that show fractal properties in spatial and temporal dimensions. The electrons are released by collective electron emission processes that are thermionic or determined by a strong electric field. Each emission center is active for a short

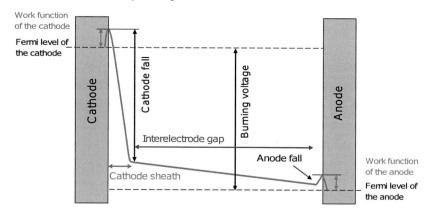

Fig. 2.2: Simplified presentation of the potential distribution between the anode and the cathode of a cathodic arc reproduced after Ref. [20]. The potential is plotted as negative potential to account for the negative charge of the electron. The sheaths are very thin and shown not to scale.

period of time, then it extinguishes and ignites at a new location on the cathode. During arc operation the life cycle of an emission center can be divided into four stages, namely (i) the pre-explosive stage, (ii) the explosive emission stage, (iii) the immediate post-explosion stage and (iv) the final cool-down stage. In the first stage a new emission site is ignited. The ignition takes place at favorable surfaces sites that are characterized by a low local work function and high electric field strength at a thin cathode sheath and a high intensity of ion bombardment. As a result of the local energy input electron emission with thermal runaway takes place at these sites leading to the next stage. During stage (ii) molten pools of liquid metal are formed and explosive electron emission occurs. Within this stage a fully ionized plasma is formed that contains multiple charge states (1+ to 5+). As a result of the microexplosions also microscopic droplets so-called macroparticles are emitted. Macroparticles are usually liquid when they form and cool down while travelling away from the cathode. If they arrive on the substrate in liquid state, they can be incorporated into the growing film. If they arrive as solid particles at the substrate, they can be bounced back.

The explosive stage is followed by a longer decay stage, the so-called immediate post-explosion stage. In that stage the liquid metal layer of the freshly created crater emits electrons and vapor. The vapor becomes ionized close to the cathode surface as a result of electron atom collisions within the intense electron beam formed in the thin cathode sheath. After all the final cool-down stage starts.

Due to the explosive nature of the cathode processes the net flux of positive ions is away from the cathode and therefore denoted as anomalous ion emission. The electrons move to the anode. In the absence of an external magnetic field, the location of the ignition and extinction of active emission sites moves randomly across the cathode, which appears like the random movement of the spot.

By the application of a magnetic field the arc or rather the motion of the ignition locations can be steered. If the magnetic field vector $\vec{B}$ is parallel to the cathode surface, the plasma column is bent in the direction $\vec{j} \times \vec{B}$ (where $\vec{j}$ is the current density) and the virtual motion of the cathode spot is to the opposite direction if the pressure is $< 10^3$ Pa. The spot motion in the direction $-\vec{j} \times \vec{B}$ is often termed as "retrograde" motion. If the magnetic field vector is tilted with respect to the cathode's surface normal, then the direction of the virtual motion is tilted to the retrograde direction by the Robson angle.

As described above, CAE produces a highly ionized plasma containing multiple charged ions of the target material. The deposition of the growing film using CAE is an energetic deposition which is determined by the ions of the film material. Compared to the flux of ions, the flux of neutrals is usually negligible during CAE depositions [21]. The total energy E_{total} brought by an ion to the substrate is described by Anders as a sum of different energy contributions according to the following equation:

$$E_{total}(Q) = \underbrace{E_{i0} + \Delta E_{i,kin} + E_{ic}}_{kinetic\ energy} + \underbrace{E_c + E_{exc} + E_{ion}}_{potential\ energy} - Qe^{\ominus}\phi_s$$

(2.4)

One energy contribution is the natural kinetic energy E_{i0} gained at the cathode spot. The ion velocity v_{i0} close to the cathode spot reaches supersonic velocity in the range $1 - 2\times10^4$ m/s which is almost independent of the cathode material and the ion charge state [21]. The kinetic energy of a cathodic arc ion with the mass m_i corresponds to Eq. (2.5) and can be modified before reaching the substrate due to collisions with neutrals and other interactions within the plasma.

$$E_{i0} = m_i v_{i0}^2 / 2$$

(2.5)

Due to the high ion velocities, the CAE process is characterized by a high natural kinetic energy E_{i0} of the ions which corresponds to approximately 20 eV for light elements and 200 eV for heavy elements [21].

Another energy contribution to the total energy is the kinetic energy $\Delta E_{i,kin}$ which the ions gain if the substrate surface is negative with respect to the plasma potential. This energy gain is determined by the potential difference (ΔV_{sheath}) between the substrate and the plasma potential (V_{pl}) according to Eq. (2.6).

$$\Delta E_{i,kin} = Qe^{\ominus}\Delta V_{sheath}$$

(2.6)

where $e^{\ominus}$ is the elementary charge of an electron and Q the charge state number of the ion. The potential of the substrate is adjusted by the substrate bias voltage (U_B) that is given with respect to ground. As the potential difference ΔV_{sheath}, which determines $\Delta E_{i,kin}$, is defined by Eq. (2.7) the application of a substrate bias voltage is used to control the energy of the arriving ions.

$$\Delta V_{sheath} = |U_B - V_{pl}|$$

(2.7)

The term E_{ic} represents the gain of additional small amount of kinetic energy by image charge acceleration when the ion approaches the surface. According to Refs. [19, 22] this can be explained as follows. If an ion approaches a surface of a solid, it induces a rearrangement of the electron density in the solid a so called "image". This image charge of opposite polarity accelerates the ion towards the solid's surface. This is called image charge acceleration and the resulting energy gain E_{ic} is the image energy gain. Further details are given in Refs. [19, 22].

The ions arriving at the substrate have also potential energy which is composed of cohesive energy (E_c), cumulative ionization energy (E_{ion}) and excitation energy of bound electrons (E_{exc}). The cohesive energy is released if the condensing particle is bonded to the substrate. The electronic excitation energy is small and can be neglected [19] whereas the cumulative ionization energy E_{ion}, that is the sum of ionization energies of each ionization step for a multiple charged ion, is the largest contribution to potential energy. Depending on the charge state and type of the ion, the ionization energy E_{ion} ranges from about 6 eV to tens of eV [18]. The total energy that each ion delivers to the substrate corresponds to the sum of all kinetic and potential energies just reduced by the product of the charge state number of the ion, elementary charge and work function (ϕ_S), which corresponds to the energy needed to capture Q electrons from the substrate.

Due to the different energy contributions (see Eq. (2.4)), which every ion delivers to the growing film, CAE is a very energetic deposition technique. An imagination of the magnitude of the different energy contributions can be obtained by Table 2.1 that is based on the "Periodic Tables of Cathode and Arc Plasma Data" given by Anders [19]. The values given in Table 2.1 are examples for Al^{2+}, Ti^{2+} and Ti^{3+} ions with the most likely velocity of 15400 m/s that were generated by vacuum arc with an average arc burning voltage of 21.3 V and 23.6 V for Ti and Al, respectively at an arc current of 300 A and 150 µs after arc triggering. The assumed ΔV_{sheath} was 50 eV. According to the examples, the most important contributions to the total energy are the initial kinetic energy E_{i0}, the energy gained due to the potential difference between plasma and substrate $\Delta E_{i,kin}$ and the potential energy in form of cumulative ionization energy E_{ion}. It is obvious that the charge state of an ion influences especially $\Delta E_{i,kin}$ and the potential energy contribution E_{ion}. Furthermore, biasing the substrate is an important parameter in order to vary the energy delivered to the substrate, since it influences ΔV_{sheath}.

Table 2.1: Energy contributions (in eV) of an Al^{2+}, Ti^{2+} and Ti^{3+} ion generated by vacuum arc. The energy term $\Delta E_{i,kin}$ was calculated for a sheath potential ΔV_{sheath} of 50 eV and 100 eV.

	ΔV_{sheath}	E_{io}	$\Delta E_{i,kin}$	E_{ic}	E_c	E_{ion}	$Qe^{\ominus}\phi_s$	E_{total}
Al^{2+}	50	33.1	100	2.1	3.39	24.8	8.4	154.99
Al^{2+}	100	33.1	200	2.1	3.39	24.8	8.4	254.99
Ti^{2+}	50	58.9	100	2.3	4.85	20.6	8.8	177.85
Ti^{2+}	100	58.9	200	2.3	4.85	20.6	8.8	277.85
Ti^{3+}	50	58.9	150	5.4	4.85	48.1	13.2	255.05
Ti^{3+}	100	58.9	300	5.4	4.85	48.1	13.2	405.05

Especially the total energy E_{total} given for the Al^{2+} and Ti^{2+} ions are important since these charge states represent the highest fraction according to the charge state distribution determined ~ 150 μs after arc triggering for arc currents in the range of 100 to 300 A as given in Ref. [19]. According to Ref. [19] the charge state distribution for Ti and Al equals: 11 % Ti^{1+}, 75 % Ti^{2+}, 14 % Ti^{3+} and 38 % Al^{1+}, 51 % Al^{2+}, 11 % Al^{3+}, respectively.

3 Ti-Al-N coatings

Today, transition metal nitrides produced by PVD methods are known as protective coatings for tools. A prominent candidate represents titanium nitride, which was one of the first coating systems that was used for wear resistant coatings deposited on tools. At the beginning TiN was deposited by chemical vapour deposition (CVD) technology, e.g. Ref. [23]. The CVD technique was used since the 1960s for the deposition of hard coatings on hard metal cutting tools [24]. Admittedly, tool steels could not be coated using the CVD technology, because the high deposition temperature (900 – 1100 °C [23, 25]) during CVD lead to overtempering and softening of the steel. By the introduction of PVD methods like ion platting, arc evaporation and sputtering, low deposition temperatures (< 550 °C) were possible which allowed the coating deposition onto tool steels below their tempering temperature. Finally in the late 1970s, the large scale production of TiN coatings that were deposited onto tools using PVD methods started [26, 27] and TiN gained the leading position as wear resistant coating deposited by PVD. Later in the late 1980s, the utilization of the binary nitrides ZrN [28, 29] and CrN [30] as protective coatings which were deposited by PVD methods started.

Although already 1972 the first report about the addition of Al to TiN sputtered films and the influence of Al addition to the resistivity was published [31], the promising role of ternary Ti-Al-N coatings as protective coatings in cutting applications was realized first 1986 when Jehn [6], Knotek [5] and Münz [7] published their results about the improved oxidation resistance, hardness and cutting performance of Ti-Al-N coatings as compared to the known TiN coatings in scientific journals. At that time TiN coatings owned the leading position as wear resistant coating deposited by PVD methods. One year before, Knotek [32] and Münz [33] could already introduce their findings at a conference. Since that time, the research of Ti-Al-N coatings went on. Nowadays, Ti-Al-N coatings are used as state-of-the-art coatings in metal cutting applications [2, 34]. Further improvements of the coating properties, like e.g. hardness, oxidation resistance and thermal stability, are aspired by the addition of elements like e.g. B [35], Hf [36], Nb [37], Si [38], Ta [35], Y [39] etc. to the Ti-Al-N based coatings or by deposition of the coatings as multilayers [35, 40]. The alloying of Ti-Al-N coatings with other elements as well as the development of new and improved investigation devices e.g. transmission electron microscopy, atom probe tomography, *in situ* X-ray diffraction etc. kept the research activities about Ti-Al-N based coatings very active up to now [41].

3.1   The Ti-Al-N system

Titanium nitride crystallizes in the face centered cubic (fcc) structure with the space group
$Fm\bar{3}m$. The thermodynamically stable phase of aluminium nitride is the wurtzitic (w) phase
with the space group $P6_3mc$. Under equilibrium conditions, less than 2 at.% of Al can be
dissolved in fcc-TiN at 1000 °C, whereas no detectable dissolution of Ti was found in w-AlN
at the same temperature [42]. Under non-equilibrium conditions that are present during coating
deposition using PVD techniques, it is possible to dissolve a considerable amount of Al into the
fcc structure of TiN by the substitution of Ti atoms with Al atoms and to form the metastable
fcc-$Ti_{1-x}Al_xN$ phase. The highest Al content in Ti-Al-N coatings that led to the formation of
fcc-$Ti_{1-x}Al_xN$ single phase coatings was $x = 0.67$. This solubility limit was reported by
Setsuhara et al. [43] who deposited $Ti_{0.33}Al_{0.67}N$ coatings by ion beam assisted deposition and
by Hörling et al. [44] who used arc evaporation. Both authors found that at higher Al contents
of $x = 0.71$ [43] and 0.75 [44] the coatings were dual phase containing cubic and wurtzitic phase.
If the Al content was further increased e.g. to $x = 0.84$ [43] or $x = 0.85$ [45], the wurtzitic phase
was observed as single phase.

A lower solubility limit of Al in fcc-$Ti_{1-x}Al_xN$ was observed e.g. by Wahlström et al. [46] and
Zhou et al. [47] who obtained single phase fcc-$Ti_{1-x}Al_xN$ coatings up to $x \leq 0.4$ using magnetron
sputtering and $x \leq 0.5$ using RF magnetron sputtering, respectively. The dual phase region
where fcc and wurtzitic structures were present was $0.42 \leq x \leq 0.9$ in case of Ref. [46] and
$0.6 \leq x \leq 0.7$ in case of Ref. [47]. The different solubility limits of Al in the metallic sublattice
of fcc-TiN that were found experimentally in PVD coatings are attributed on the one hand to
the great variety of PVD techniques and the choice of the deposition parameters like substrate
bias voltage and substrate temperature. On the other hand the precision in the determination of
the phase composition might affect the reported results because the early formation stages of
the wurtzitic phase might be not recognized very easily [48]. The reasons for this are diverse.
On the one hand AlN has a lower scattering power then TiN due to the lower atomic scattering
factor of aluminium as compared to titanium. Thus, the diffracted intensity of the same volume
of w-AlN is lower than for fcc-TiN which hampers the identification of wurtzite phase
especially if minor amounts of the wurtzite phase are present. On the other hand microstructure
defects lead to broadening of the diffraction lines of w-AlN which makes their identification
more difficult. In the case of stacking faults that are distributed on the basal planes the
broadening of the diffraction lines is anisotropic [49]. Thus, the diffraction lines with $h - k =
3n \pm 1$ (101_w, 102_w, 103_w and 202_w) are strongly broadened, whereas other diffraction lines are
unaffected [48, 50]. Furthermore, the epitaxial growth of the wurtzite phase on fcc-(Ti,Al)N

leads to a stronger overlap of the diffraction lines from wurtzite phase and fcc-(Ti,Al)N which limits the visibility of the diffraction lines of the wurtzite phase [48]. *Ab initio* calculations by Mayrhofer *et al.* [51] and Zhang *et al.* [52] support the relative high Al solubility limit in metastable fcc-$Ti_{1-x}Al_xN$ of $x = 0.67$ mentioned above. According to the calculated Gibbs free energy diagram of the immiscible quasi binary TiN-AlN system for fcc and wurtzite structures in Ref. [52] the fcc structure is more stable for $x < 0.68$ and the wurtzite structure is more stable for $x > 0.68$ as calculated for $T = 27$ K. Mayrhofer *et al.* [51] calculated the influence of the number of Ti-Al bonds on the solubility limit using a supercell consisting of 32 and 64 metal atoms. The comparison of the calculated energy of formation of w-$Ti_{1-x}Al_xN$ and fcc-$Ti_{1-x}Al_xN$ for different numbers of Ti-Al bonds per unit cell implied that the decreasing number of Ti-Al bonds increases the Al solubility in fcc-$Ti_{1-x}Al_xN$ from $x = 0.64$ to $x = 0.74$. The authors of Ref. [51] explain this by a lower contribution of the configurational entropy if the number of Ti-Al bonds is high which leads to a higher energy of formation and thus to a lower solubility limit.

3.2 Review of Ti-Al-N coatings deposited by CAE

The properties of Ti-Al-N coatings are determined by microstructure features like phase composition, crystallite size and lattice strain. Different parameters of the deposition process (e.g. ion to neutral flux, temperature, substrate bias voltage, gas pressure, etc.) influence the microstructure of the deposited coatings. In case of Ti-Al-N coatings deposited by CAE, the substrate bias voltage and the Al content play a crucial role in the adjustment of the microstructure. Most publications about CAE $Ti_{1-x}Al_xN$ coatings report either on the influence of the Al content (see Chapter 3.2.1) or the bias voltage (see Chapter 3.2.2) on phase composition, hardness and residual stress. A detailed study of $Ti_{1-x}Al_xN$ coatings deposited by CAE that comprises the influence of different Al content and bias voltages on the evaluation of macroscopic stress, phase composition, crystallite size and lattice parameter is not available in literature.

3.2.1 Influence of the Al content on the microstructure and properties of Ti-Al-N coatings deposited by CAE

The influence of an increasing Al content on the phase composition, hardness as well as lattice parameter and residual stress of the fcc-(Ti,Al)N phase in $Ti_{1-x}Al_xN$ coatings, that were deposited by CAE, was studied by Ikeda *et al.*, Kimura *et al.*, Santana *et al.*, Rafaja *et al.* and Hörling *et al.* [45, 53-57]. The results of these investigations are summarized in Fig. 3.1.

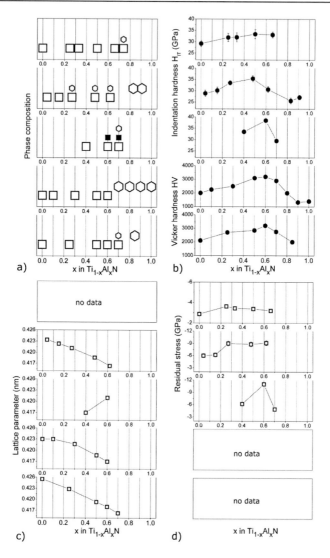

Fig. 3.1: Influence of the Al content on the phase composition (a), hardness (b), lattice parameter of the fcc-(Ti,Al)N phase (c) and residual stress of the fcc-(Ti,Al)N phase (d) in $Ti_{1-x}Al_xN$ coatings that were deposited by CAE. The results shown above were obtained by the research groups of Ikeda *et al.* [45], Kimura *et al.* [53], Santana *et al.* [54], Rafaja *et al.* [55, 57] and Hörling *et al.* [56]. The order of the references corresponds to the sub-graphs from the bottom to the top. The symbols in figure (a) correspond to fcc-(Ti,Al)N (□), fcc-AlN (■) and wurtzitic (hexagon) phase. The lines in figures (b), (c) and (d) are guide for the eyes.

Although the deposition parameters used by the different research groups were not identical with respect to substrate bias, temperature and nitrogen pressure, similar trends in the evolution of the microstructure parameters are visible in Fig. 3.1. All authors observed metastable fcc-(Ti,Al)N as single phase at low Al contents (see Fig. 3.1a). Santana *et al.* [54] found fcc-AlN for $x \geq 0.6$ as second phase. The presence of wurtzite phase as additional phase was found by most of the authors [45, 54, 56] at $x \geq 0.7$. In contrast to that, Rafaja *et al.* [55, 57] deduced the presence of wurtzite phase as second phase in addition to fcc-(Ti,Al)N already at $x \geq 0.28$. At high Al contents wurtzite phase was identified as single phase by Kimura *et al.* [53] at $x \geq 0.7$ and by Ikeda *et al.* [45] and Rafaja *et al.* [55] for Al contents exceeding $x = 0.8$.

All authors [45, 53, 54, 56, 57] observed an increase of the hardness with increasing Al content up to a certain Al concentration followed by a hardness reduction (Fig. 3.1b). Rafaja *et al.* [58] could correlate the hardness maximum with the dual phase character of the $Ti_{1-x}Al_xN$ coatings when wurtzite and fcc phase were present. In case of the other authors [45, 53, 54, 56] the region with the highest hardness coincided with the highest Al content at which no wurtzite phase was recognized.

The lattice parameters for fcc-(Ti,Al)N as function of the Al content in the coatings as presented by the authors of References [45, 53-55] are shown in Fig. 3.1c. The absolute values have to be considered carefully since the values given in [45, 53, 54] might be affected by residual stress, because the authors did not indicate whether the stress-free lattice parameter was determined. However, a reduction of the lattice parameter with increasing Al content was observed in References [45, 53, 55] which is attributed to the smaller Al atoms substituting Ti atoms in the fcc structure. The increase of the lattice parameter in Ref. [54] from $a = 0.417$ nm for $x = 0.4$ to $a = 0.421$ nm for $x = 0.6$ attribute the authors to a reduced incorporation of Al in fcc-(Ti,Al)N and the formation of cubic AlN.

The residual stress of the fcc-(Ti,Al)N phase evaluated by References [54-56] and shown in Fig. 3.1d was determined by X-ray diffraction. The absolute values of the residual stress obtained by the different research groups cannot be compared directly, because deposition parameters (e.g. applied bias voltage, substrate material and total coating thickness), which influence the stress, as well as the elastic modulus and Poisson's ratio, that were used for the calculation of the residual stress (cf. Eq. (6.24)), were not equal for all sample series. Nevertheless, similar trends in the evolution of the residual stress of the fcc-(Ti,Al)N phase with increasing Al content were visible, because all authors observed compressive stress in the fcc-(Ti,Al)N phase and according to the results obtained by [55, 56] the addition of Al yielded in a slight increase of the compressive stress as compared to nearly Al-free films.

3.2.2   Influence of the bias voltage on the microstructure and properties of Ti-Al-N coatings deposited by CAE

Only a limited number of authors studied the influence of the bias voltage (U_B) on the microstructure and properties of Ti-Al-N coatings that were deposited by CAE [59-62]. The coatings studied under this aspect were confined just to two aspired coating compositions, because these coatings were deposited from mixed Ti-Al-targets with the composition $Ti_{0.5}Al_{0.5}$ [59, 60, 62] or $Ti_{0.4}Al_{0.6}$ [61]. All authors dealing with the influence of the bias voltage on Ti-Al-N coatings presented its effect on the coating composition and residual stress of the fcc-(Ti,Al)N phase. Since the calculation of the residual stress of fcc-(Ti,Al)N requires its lattice parameter (see Section 6.2.1.1), the authors determined experimentally the lattice parameter of fcc-(Ti,Al)N but apart from Ref. [59] it is not clear if the determined lattice parameter is corrected with respect to stress. The hardness, Young's modulus and adhesion performance during drilling or turning tests as function of the bias voltage were shown in References [60-62]. Among the Refs. [60-62], the most comprehensive study was done by Sato *et al.* [61] who evaluated additionally the line broadening of the 111, 200 and 220 diffraction line of the fcc-(Ti,Al)N phase and evaluated the column size from TEM cross sections. The results obtained by Oettel *et al.*[59], Vlasveld *et al.* [60], Ahlgren *et al.* [62] and Sato *et al.* [61] are summarized in Fig. 3.2 with regard to the actual Al content in the coatings, the hardness, the coating thickness as well as the residual stress and lattice parameter of the fcc-(Ti,Al)N phase. A direct comparison of the results obtained by the different research groups is very difficult due to different substrate material, temperatures and nitrogen pressure during deposition as well as different or even unknown values for Young's modulus and Poisson's ratio, that influence the determination of the residual stress (see Section 6.2.1.1). However, some trends in the evolution of the microstructure parameters and hardness can be identified in the following.

3.2.2.1   Influence of U_B on the Al content in the Ti-Al-N coatings deposited by CAE

All authors [59-62] observed that the [Al] / ([Al] + [Ti]) ratio in the coatings is lower than in the mixed Ti-Al-targets which were used for the deposition. The reduction of the Al content in the coatings as compared to the respective target composition is shown in Fig. 3.2a. The degree of the observed Al reduction is very different, e.g. for a substrate bias voltage of -100 V the drop of the Al fraction in the coatings ranged from ~ 5 % in Ref. [62] to ~ 31 % in Ref. [59]. The authors of Ref. [59-61] observed an increased Al reduction in the coatings with increasing bias voltage which was very high in the study of Oettel *et al.* [59] and moderate at the center of the sample in the study of Sato *et al.* [61] (see Fig. Fig. 3.2a).

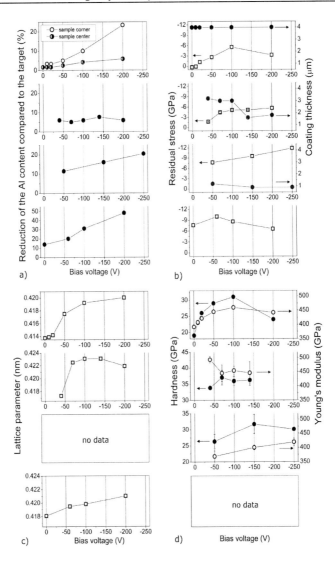

Fig. 3.2: Influence of the substrate bias voltage on the Al reduction in the coating (a), residual stress of the fcc-(Ti,Al)N phase and coating thickness (b), lattice parameter of the fcc-(Ti,Al)N phase (c) and hardness (d) in Ti$_{1-x}$Al$_x$N coatings that were deposited by CAE. The results shown above were obtained by the research groups of Oettel *et al.* [59], Vlasveld *et al.* [60], Ahlgren *et al.* [62] and Sato *et al.* [61]. The order of the references corresponds to the sub-graphs from the bottom to the top.

In contrast to that, Ahlgren *et al.* [62] observed a relatively constant Al reduction ranging from 5 to 8 % in the U_B range from -40 V to -200 V. Furthermore, Sato *et al.* [61] found that the Al reduction at the corner of the sample is intensified as compared to the center of the sample. The Al reduction in the coatings during CAE deposition is due to a higher resputtering of already deposited Al on the substrate. This is attributed to a higher sputter yield of Al in comparison to Ti. The sputter yield can be calculated with the SRIM-Program [63]. In case of a sheath potential ΔV_{sheath} of 100 V (that could be reached by a bias voltage of approx. -100 V), a reasonable energy of an Al^{2+} ion generated by the cathodic arc would be ~ 250 eV (see Table 2.1). If such an ion bombards an Al and Ti surface perpendicularly, then the sputter yield equals 0.68 for Al and 0.51 for Ti as calculated by the *SRIM Program* [63]. In case of a Ti^{2+} ion with an energy of 280 eV (see Table 2.1) the sputter yield for Al corresponds to 0.37 and for Ti to 0.31 [63]. Thus the number of resputtered Al is higher in comparison to Ti. The resputtering effect is pronounced at the corner edge since more ions were focused during deposition [61] leading to a distinct Al loss (see Fig. 3.2a). Furthermore, an increased ionization of Ti over Al takes places during CAE. The charge state distribution of single, double and triple charged ions in a vacuum arc were 11 %, 75 % and 14 % for Ti and 38 %, 51 % and 11 % for Al, respectively [64]. This results in a mean charge state of 2.1 for Ti and 1.7 for Al. Thus, at a certain bias voltage the Ti ions gain in average more kinetic energy $\Delta E_{i,kin}$ according to Eq. (2.6) than Al ions. Consequently, Ti ions can be implanted deeper in the coating. This in turn requires a higher energy in order to release them from the deposited coating surface [65].

3.2.2.2 Influence of U_B on the residual stress in Ti-Al-N coatings deposited by CAE

The authors of Ref. [59-62] determined the macroscopic residual stress of the fcc-(Ti,Al)N phase by X-ray diffraction using either the $\sin^2 \psi$ method or modified $\sin^2 \psi$ method (see Section 6.2.1.1). The results of the stress analysis are summarized in Fig. 3.2b.

According to Oettel *et al.* [59] the macroscopic residual stress (σ) comprises three different stress contributions and is described by Eq. (3.1):

$$\sigma = \sigma_g + \sigma_{th} + \sigma_p \tag{3.1}$$

The first contribution originates from the growing process (σ_g), the second one stems from the difference in the thermal expansion of the layer (α_L) and the substrate (α_S), that is called thermal stresses (σ_{th}) and the third one is caused by phase transformation or precipitation processes during cooling (σ_p). The thermal stresses present at the measuring temperature ($T_{measure}$) can be calculated according to Eq. (3.2) [59]:

$$\sigma_{th} = -(\alpha_S - \alpha_L) \cdot (T_S - T_{measure}) \frac{E}{1 - \nu} \qquad (3.2)$$

where T_S is the substrate temperature during deposition and E and ν are the Young's modulus and Poisson's ratio of the coating, respectively.

Oettel *et al.* [59] attributes the grown-in stresses σ_g to the generation of lattice defects. As the bias voltage determines the energy of the incoming ions (see Eq. (2.6)), it is considered as the main factor influencing the generation of lattice defects. In several publications the macroscopic residual stresses present in the coatings are attributed to the formation of lattice defects [66]. However, the kind of these defects are often not explained. In case of nitride layers, Oettel *et al.* [59] considers metal and nitrogen atoms on irregular lattice sites and nitrogen on interstitial sites as the source of residual stresses.

Sato *et al.* [61] calculated the thermal stress component (σ_{th}) for Ti$_{0.4}$Al$_{0.6}$N coatings with $\alpha_L = 7.2 \cdot 10^{-6}$ K deposited on cemented carbide substrates made of WC-Co with $\alpha_S = 5.5 \cdot 10^{-6}$ K for a deposition temperature of 450°C. In that case the stresses generated by the different thermal expansion coefficients are tensile and correspond to 0.38 GPa. This stress level coincides with the macroscopic stress determined by Sato *et al.* [61] for the coatings deposited at $U_B = 0$ V which was 0.32 GPa. At this bias voltage the generated ions are not accelerated by the bias voltage and cause no significant intrinsic stresses. In contrast to that with increasing U_B to -100 V, the compressive stress increases which the authors of Ref. [61] attribute to the high energy ion bombardment. A further increase of U_B to -200 V yielded in a lower compressive stress than at -100 V which the authors explain by the reduced incorporation of Al in the coating (cf. Fig. 3.2a) and an increased mobility of the atoms due to raised local deposition temperature at high U_B [61]. Since the coating thickness of all coatings in Ref. [61] was constant, an effect of the thickness on the evolution of the macroscopic residual stress can be excluded.

Ahlgren *et al.* [62] also observed an increase of the compressive stress when the negative bias voltage was increased from 40 V to 70 V. At higher negative bias voltages the compressive stress saturated. The coating thickness was lower at $U_B = -150$ V and -200 V which could affect the stress level in the coating. Generally, it is expected that the stresses are reduced with increasing coating thickness due to longer processing times that promote the mobility of the defects and their self-annealing during the deposition process [59].

Vlasveld *et al.* [60] observed the continuous increase of the residual stress with increasing U_B from -50 V to -250 V (see Fig. 3.2b) which he attributes to higher defect concentrations as a result of the shot peening process due to the increased energy of the bombarding ions with increasing U_B. The evaluation of the stress is not influenced by the coating thickness since it is relative constant ranging from 0.9 to 1.2 μm.

Oettel *et al.* [59] found for their coatings a high compressive stress of at least 6 GPa. The coatings were deposited on high speed steel substrates having a thermal expansion coefficient of $\alpha_S = 10 \cdot 10^{-6}$ K [67] which is higher than the one for $Ti_{0.5}Al_{0.5}N$. Thus, the stress contribution from the difference in thermal expansion coefficients is compressive and lower than 0.5 GPa [59]. Oettel *et al.* [59] attributes the high compressive stress present in his coatings to ion bombardment of the growing film. In contrast to Sato *et al.* [61], Oettel *et al.* [59] found a high stress level also if no bias voltage was applied. According to him, the ions have a high energy which they collect in the electric field of the arc so that ions deliver a high energy also without biasing of the substrate. In his opinion, metal ions with an energy of several hundred electron volts punch nitrogen atoms into irregular lattice sites and occupy the previous nitrogen sites since the metal ions (e.g. $r_{Ti}^{2+} = 0.76$ nm) have a similar radius like the nitrogen atoms ($r_N = 0.70$ nm) [59]. The irregular nitrogen stays at the interstitial lattice sites or diffuses to the surface. According to Oettel's idea [59], the nitrogen at interstitials as well as the metal atoms on nitrogen sites cause the high grown-in compressive stresses σ_g. An experimental evidence for his assumption is not shown in Ref. [59]. With increasing bias voltage a rise of the substrate temperature from 250 to 320°C was observed which is considered to promote the annihilation of ion induced defects [59]. As a result of this, the authors of Ref. [59] consider the evolution of residual stress vs. bias voltage as independent.

The main conclusions from the results shown above can be summarized as follows: The ion induced defects are considered by most authors [59-62] as the main reason for the compressive stress state of fcc-(Ti,Al)N in Ti-Al-N-coatings deposited by CAE. The ion induced defects arise from a shallow subsurface implantation the so called "subplantation" [68] of the ions with an energy exceeding the subplantation threshold [69]. Furthermore, most authors observed a saturation of the compressive stresses with increasing substrate bias voltage which they attribute to defect annihilation due to increased substrate temperature as a result of the ion bombardment. The thermal stresses are either compressive or tensile depending on the substrate material and their contribution to the total stress value is low. All authors mentioned above assumed that the investigated Ti-Al-N-coatings contained fcc-(Ti,Al)N as single phase. A possible contribution of a second phase on the stress level in the Ti-Al-N-coatings deposited by CAE is not discussed in References [59-62].

3.2.2.3   Influence of U_B on lattice parameter of fcc-(Ti,Al)N in CAE coatings

The lattice parameters of fcc-(Ti,Al)N as a function of U_B were determined experimentally by the authors of References [59, 61, 62] and are summarized in Fig. 3.2c. Apart from Ref. [59] it is not announced whether the lattice parameter was corrected with respect to stress. Hence, a comparison of the lattice parameter is not reasonable. The increase of the stress-corrected lattice parameter with increasing bias voltage observed by Oettel *et al.* [59] was attributed to the reduced Al incorporation into the Ti-Al-N coatings.

3.2.2.4   Influence of U_B on the hardness of Ti-Al-N coatings deposited by CAE

The hardness evolution of Ti-Al-N coatings with increasing U_B observed by the authors of Refs. [60-62] is summarized in Fig. 3.2d. Sato *et al.* [61] and Ahlgren *et al.* [62] found that the trend in the evolution of the compressive stress vs. U_B is symmetrical to the trend in the hardness evolution vs. U_B. In case of the experiments done by Vlasveld *et al.* [60] the hardness evolution vs. U_B deviated from the evolution of the compressive stress because at the highest bias voltage the hardness saturated whereas the compressive stress increased further. This effect could not be explained by the authors. A possible contribution of the crystallite size to the hardness evolution was considered by none of the authors of References [60-62].

3.3   Thermal stability of Ti-Al-N coatings

Until 2010 the thermal stability of Ti-Al-N coatings deposited by magnetron sputtering [70-72] and CAE [40, 44, 54, 56, 73-76] was investigated in different studies. The great interest in this topic is proven by a further increase in the number of publications within the years 2011 to 2013 since References [36, 39, 77-81] and References [78, 82-89] focus on the thermal stability of Ti-Al-N coatings deposited by magnetron sputtering and CAE, respectively.

Already 1991, Adibi *et al.* [70] found that surface-initiated spinodal decomposition takes places during film growth of magnetron sputtered $Ti_{0.5}Al_{0.5}N$ coatings at 540 - 560°C resulting in the formation of coherent fcc-TiN-rich and fcc-AlN-rich platelets. After the increase of the deposition temperature to $T > 560°C$ wurtzitic AlN was formed incoherent with the composition modulated cubic matrix. At growth temperatures of $T > 750°C$ the coatings consisted of w-AlN and fcc-TiN.

More than 10 years after Adibi's contribution [70], the decomposition sequence during annealing of single-phase fcc-$Ti_{0.33}Al_{0.67}N$ [44, 73] and fcc-$Ti_{0.5}Al_{0.5}N$ [73] coatings was described by Hörling *et al.* [44] and Mayrhofer *et al.* [73]. According to them the decomposition of metastable fcc-(Ti,Al)N occurs into two steps. In the first step the fcc-$Ti_{1-x}Al_xN$ coatings undergo spinodal decomposition and form coherent fcc-TiN-rich and fcc-AlN-rich domains. The start of this process was observed in the temperature range ~ 860-900°C.

At higher temperatures the next decomposition step takes place in which fcc-AlN transforms into its thermodynamically stable wurtzite phase by nucleation and growth. At the end of the decomposition process the coatings consist of fcc-TiN and w-AlN. Indications for the formation of w-AlN were found by Hörling *et al.* [44] after annealing a $Ti_{0.33}Al_{0.67}N$ coating at 1100°C. At this temperature w-AlN coexisted with fcc-AlN and Al-depleted fcc-(Ti,Al)N matrix. In contrast to this, Mayrhofer *et al.* [73] found only the cubic structures of AlN and TiN in the fcc-(Ti,Al)N matrix at 1100°C and 1030°C in $Ti_{0.33}Al_{0.67}N$ and $Ti_{0.5}Al_{0.5}N$ coatings, respectively. The end of the phase transformation of fcc-AlN into w-AlN was observed by Hörling *et al.* [44] at 1250°C and by Mayrhofer *et al.* [73] at 1400°C. The isostructural decomposition into fcc-TiN-rich and fcc-AlN-rich domains is associated with an increase of hardness as shown in Ref. [73] for $Ti_{0.33}Al_{0.67}N$ coatings. Meanwhile this "age hardening" effect at ~ 900°C was shown also for $Ti_{1-x}Al_xN$ coatings with Al contents in the range of $0.48 \leq x \leq 0.59$ [36, 39, 77, 78]. The age hardening effect of $Ti_{1-x}Al_xN$ coatings at this temperature range is beneficial for their application as coatings for cutting tool inserts. Because recently it was shown that the temperature at the cutting edge of $Ti_{0.6}Al_{0.4}N$ coated WC-Co cutting inserts was 850-900°C during dry cutting of carbon steel [88].

The formation of w-AlN was frequently reported to take place after annealing at temperatures above 1000°C [36, 39, 40, 44, 73, 74, 76, 79, 87]. Moreover for a long time, the formation of w-AlN was considered to be detrimental for the coating properties like hardness e.g. References [45, 47, 90] and adhesion [56, 66]. However, in the recent time some publications showed that the presence of w-AlN is not generally harmful with regard to the hardness of Ti-Al-N coatings. Their hardness is rather determined by the amount of w-AlN and the manner of its incorporation into the coatings as it was shown for as-deposited Ti-Al-N coatings containing w-AlN [38, 57, 91-93]. In addition, recent publications about the thermal stability of Ti-Al-N coatings showed that w-AlN can be already formed during annealing in the temperature range of 850-900°C [78, 80-84, 94]. Exactly after annealing at 900°C a hardness maximum was observed in References [80, 94] when a minor amount of w-AlN was formed in addition to fcc-TiN-rich and fcc-AlN-rich domains. Thus, it could be evidenced that w-AlN, which forms during annealing, is not mandatory negative for the hardness evolution. It rather implies that the so called "age hardening" of Ti-Al-N coatings is not only caused by isostructural decomposition into fcc-TiN-rich and fcc-AlN-rich domains but also by the formation of a small amount of w-AlN, whose presence might be not recognized since the early formation stages of the wurtzitic phase might be not identified easily [48] (see also Section 3.1 page 12).

4 Ti-Al-Ru-N coatings

In order to improve the properties of Ti-Al-N coatings and their thermal stability the influence of different alloying and doping elements is discussed in literature. The most popular elements were Y [95, 96], Si [97] and transition metals like Ta [98], V [99], Zr [100], Nb [37], Hf [37]. The addition of up to 5 at.% Ru as doping element to Ti-Al-N coatings is protected by a patent [101]. In this patent the cutting performance of Ti-Al-N / Al-Ti-(Ru)-N multilayer coatings deposited from $Ti_{50}Al_{50}$ and $Ti_{33-y}Al_{67}Ru_y$ targets with $y = 0$, 1 and 5 were compared. The coatings with an individual layer thickness of 7 nm and a total thickness of ~ 5 μm were deposited on cemented carbide cutting inserts. The Ru containing multilayers were characterized by an increased lifetime as compared to the Ru-free multilayers during turning of steel with cooling lubricant as well as dry milling of steel. Apart from one publication about Ti-Al-Ru-N monolayer coatings [102] the microstructure of Ru containing Ti-Al-N coatings is not published so far by other research groups. The state of knowledge about these coatings is presented in the next Sections 4.1 and 4.2.

4.1 Properties of Ti-Al-Ru-N coatings deposited by CAE

Pohler *et al.* [102] compare the phase composition, hardness and tribological properties of an Ru-free Al-rich Ti-Al-N monolayer and Al-rich Ti-Al-Ru-N monolayers with the addition of two different Ru concentrations that were produced from $Ti_{33}Al_{67}$ and $Ti_{32}Al_{67}Ru_1$ and $Ti_{28}Al_{67}Ru_5$ targets, respectively. The deposition was done at a bias voltage of -40 V and a temperature of 450°C using CAE. The authors expect, that Ru is present in its metallic form in the Ti-Al-Ru-N coatings because it is reported that fcc-RuN decomposes above 100°C [103]. The formation of nitrides and intermetallic phases in Ti-Al-Ru-N coatings was also excluded by Ref. [101]. The as-deposited $Ti_{0.38}Al_{0.62}N$, $Ti_{0.36}Al_{0.63}Ru_{0.01}N$ and $Ti_{0.32}Al_{0.63}Ru_{0.05}N$ coatings were dual phase containing fcc-(Ti,Al)N and a predominant w-(Al,Ti)N phase. With increasing Ru content in the coatings a higher amount of wurtzite phase was observed. This effect is attributed to an elevated [Al] / ([Al]+[Ti]) ratio in the coatings with increasing Ru content due to the addition of Ru at the expense of Ti in the $Ti_{33-y}Al_{67}Ru_y$ targets which shifts the Al-ratio in the films closer to the solubility limit of Al in fcc-(Ti,Al)N (see Section 3.1) [43, 44]. According to Ref. [102], the hardness of the as-deposited coatings decreases slightly with the addition of Ru which is ascribed to the higher amount of wurtzitic phase in the Ru-rich coating. The abrasive wear resistance at low temperatures was positively influenced by the addition of a high Ru content. In contrast to that, tribological tests at 500°C and 700°C revealed a more intensive wear mechanism with increasing Ru content.

4.2   Thermal stability of Ti-Al-Ru-N coatings deposited by CAE

Annealing of the Ti-Al-(Ru)-N coatings from Section 4.1 in vacuum revealed a retarded decomposition of fcc-(Ti,Al)N with the addition of Ru [102]. Annealing at 800°C did not cause a change in the phase composition of the as-deposited coatings. The transformation of metastable fcc-(Ti,Al)N into fcc-TiN and fcc-AlN as well as the depletion of Ti from w-(Ti,Al)N occurred between 900 C and 1000°C. The authors concluded from the comparison of the diffraction patterns obtained after annealing a retarded decomposition process in the Ru containing coatings.

5   Cr-Al-Si-N coatings

After the successful introduction of Ti-Al-N PVD films as wear resistant coatings especially for cutting tools in the 1980s (see Section 3, page 11), Cr-Al-N films deposited by PVD were introduced as promising hard coatings at the beginning of the 1990s [8, 9]. Cr-Al-N coatings are characterized by a good corrosion resistance against fluids like cooling lubricants as well as oxidation and wear resistance [4, 34, 104-106]. Due to these properties, Cr-Al-N films are used as protective coatings in many industrial applications, e.g. for metal forming dies [3], punching of perforated sheets [4] and high-end spindle bearings [107].

In order to increase the hardness and the oxidation resistance of Cr-Al-N based coatings, silicon is added as doping element [108-112]. Nowadays, Cr-Al-Si-N coatings are used successfully in industrial applications [2].

5.1   The Cr-Al-(Si)-N system

CrN crystallizes in the face centred cubic (fcc) structure with the space group $Fm\bar{3}m$. Using PVD methods the metastable fcc-$Cr_{1-x}Al_xN$ phase can be formed. The experimentally observed solubility limit of Al in the fcc- $Cr_{1-x}Al_xN$ phase lies in the range of x = 0.6 to x = 0.7 [113-117]. Aluminium contents exceeding the solubility limit led to the formation of the wurtzite phase. Theoretical predictions about the solubility limit of Al in fcc-$Cr_{1-x}Al_xN$ yielded x = 0.772 using the two-band parameter method [118] and x = 0.815 at 0 K using first principles ab initio calculations based on the local density approximation of DFT [119].

The increasing Al incorporation into fcc-$Cr_{1-x}Al_xN$ leads to the shrinkage of the fcc elementary cell, which was observed for instance in Refs. [113, 115, 116, 120-122]. The decrease of the lattice parameter with increasing Al content in the fcc-$Cr_{1-x}Al_xN$ phase present in Cr-Al-N coatings deposited by CAE was described in Ref. [92] as

$$a(Cr_{1-x}Al_xN) = [0.41486(2) - 0.00827(1) \cdot x]nm \tag{5.1}$$

As mentioned above, Cr-Al-N coatings can be doped by Si in order to further improve the hardness and oxidation behaviour.

Regarding other transition metal nitrides systems that are doped with Si like Nb-Si-N, Zr-Si-N, Cr-Si-N and Ti-Si-N, on the one hand the partial incorporation of Si into the fcc-phase as well as the formation of amorphous Si_3N_4 were observed in Cr-Si-N [123, 124], Nb-Si-N and Zr-Si-N [125]. On the other hand, the incorporation of Si into the fcc-phase was negated and instead the direct formation of amorphous Si_3N_4 or SiN_x was observed in case of Ti-Si-N [126, 127].

In case of Cr-Al-(Si)-N coatings that were deposited by CAE a partial incorporation of Si into fcc-(Cr,Al,Si)N was observed [108, 128]. The comparison of the lattice parameter of the fcc phase observed in Cr-Al-(Si)-N coatings with and without Si addition, which were prepared by CAE, indicated the inflation of the lattice parameter due to the incorporation of Si into the fcc phase which was described by the dependence given in Eq. (5.2) [109].

$$a(Cr_{1-x-y}Al_xSi_yN) = [0.41486(2) - 0.00827(1) \cdot x + 0.034(1) \cdot y]nm \qquad (5.2)$$

The incorporation of Si into fcc-(Cr,Al,Si)N was observed at low Si concentrations [10, 84, 108]. If the Si content exceeds a critical value, amorphous SiN_x starts to segregate at microstructure defects like crystallite boundaries according to the model proposed by Sandu *et al.* [125]. Indications for the start of this process were observed in Ref. [10] for an [Si] / ([Si] + [Al] + [Si]) ratio of 0.05 in Cr-Al-Si-N coatings deposited by CAE. At an [Si] / ([Si] + [Al] + [Si]) ratio of 0.08 a continuous amorphous layer surrounded the crystallites [10]. The segregation of amorphous SiN_z or rather amorphous Si_3N_4 leads to the formation of nanocrystallites that are embedded in an amorphous matrix forming so called nanocrystalline composites [127].

Thus, in the Cr-Al-Si-N coatings, the formation of nanocrystallites embedded in an amorphous matrix leads to a hardness increase in analogy to the Hall-Petch relationship [129-131]. Additionally, a solid-solution hardening at low Si concentrations due to the incorporation of Si into fcc-(Cr,Al,Si)N is assumed [108].

However, if the volume fraction of the amorphous phases exceeds a critical value with increasing Si concentration in the coatings, the hardness of the coatings decreases due to grain boundary sliding [131].

5.2   Magnetic ordering in CrN

Bulk CrN is known to be antiferromagnetic below a Néel temperature of 280 - 286 K [132-134]. The transition from the paramagnetic state to the antiferromagnetic state is accompanied by a change of the crystal structure from face centred cubic ($Fm\overline{3}m$) to orthorhombic (Pnma) [133]. Antiferromagnetism was also observed in CrN thin films [135, 136], but the antiferromagnetic ordering was smeared out in comparison with the bulk material [134].

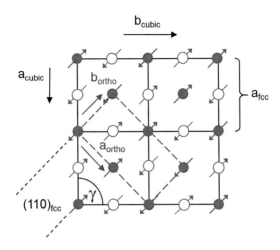

Fig. 5.1: Possible arrangement of the magnetic moments of the Cr atoms for the antiferromagnetic CrN and relationship between the cubic and orthorhombic axes, reproduced after Corliss *et al.* [132]. The solid circles correspond to Cr atoms at $z = 0$ and the open circles to Cr atoms at $z = 1/2$.

According to Corliss *et al.* [132] the transition from the cubic structure to the orthorhombic structure can be explained by Fig. 5.1 which also illustrates the relationship between the cubic and orthorhombic axes. During the transition to the orthorhombic structure, a shear distortion of the cubic cell takes places [132]. As a result of this distortion the angle γ (see Fig. 5.1) contracts to a $\sim 2°$ smaller value [133]. The lattice parameters of the orthorhombic cell given by Corliss *et al.* [132] are $a = 0.5757$ nm, $b = 0.2964$ nm $c = 0.4134$ nm. They are related to the cubic structure by $a \approx a_{fcc}\sqrt{2}$, $b \approx a_{fcc}/\sqrt{2}$, $c \approx a_{fcc}$ [133]. In case of a nanocrystalline coating with severely broadened diffraction indices, it is expected that the orthorhombic structure appears pseudo-cubic for XRD [137]. According to Corliss *et al.* [132], a ferromagnetic coupling within double layers that are parallel to $(110)_{fcc}$ planes is present in the antiferromagnetic (AFM) state. The next double layers parallel to the $(110)_{fcc}$ planes show an opposite orientation of the magnetic moments with respect to the adjacent double layers. Thus, the whole system is antiferromagnetic. This ordering is often called AFM-$[110]_2$ e.g. Refs. [138, 139]. In literature also other antiferromagnetic orderings were regarded, e.g. AFM-$[110]_1$, AFM-$[111]_1$, AFM-$[001]_1$ [133, 138-140]. It was shown, that the AFM-$[110]_1$ ordering is the most stable configuration, if the structure is constrained to cubic symmetry [133]. The AFM-$[110]_1$ ordering is characterized by ferromagnetic coupling of the spins within a single layer being parallel to $(110)_{fcc}$ planes and reverse spin directions in the adjacent single layer being parallel to $(110)_{fcc}$ planes. If planar distortion is allowed, the AFM-$[110]_2$ configuration becomes more stable.

Ab initio calculations have shown that the size of the elementary cell of CrN correlates with the magnetic ordering especially if the non magnetic, ferromagnetic and anti-ferromagnetic state are considered [138, 141]. The calculations revealed a lower lattice parameter for the non magnetic state [141].

According to Filippetti *et al.* [139] the transition from the paramagnetic to the ferromagnetic ordering or from the paramagnetic to the antiferromagnetic ordering in the fcc structure is accompanied by a volume expansion leading to an increase of the lattice parameter due to magnetic stress. In that case, the lattice parameters for the ferromagnetic and antiferromagnetic ordering were similar. However, Filippetti's calculation predicted a lower energy for the antiferromagnetic ordering than for the ferromagnetic one [139].

6 Characterisation techniques

6.1 Chemical Analysis

6.1.1 Electron probe microanalysis

Electron probe microanalysis with wavelength-dispersive X-ray spectroscopy (EPMA/WDS) was used to analyse the chemical composition of the as-deposited coatings. The EPMA/WDS measurements were done on a JXA 8900 RL from JEOL.

In the standard EPMA/WDS measurement routine the chemical composition was determined at 40 points with a distance of 20 μm on the coating surface. The acceleration voltage was set in the range of 12 to 20 kV and the beam current was 40 nA. The quantification of the measured elements was done by comparing the emitted intensity of the characteristic X-rays of the respective element with the intensity of a standard sample with known composition. The intensities were corrected with respect to dead time, drift of the beam current and background. The so called k-values that represent the ratio of the intensity of the element in the sample to the standard's intensity, were finally corrected regarding their mass, absorption and fluorescence.

The elements included in the measurement were the elements of the respective coating (e.g. Cr, Al, Si and N in case of Cr-Al-Si-N layers), oxygen and tungsten as an element of the cemented carbide substrate. The measurement of the elements of the substrate revealed that a coating thickness of at least 3 μm prevented the excitation of the electrons of the substrate elements by the incident electron beam.

In the case of Ti-Al-N and Ti-Al-N / Al-Ti-Ru-N coatings, the nitrogen content could not be measured directly due to the overlap of the spectral line $K\alpha 1$ from nitrogen ($K\alpha_1$=392.4 eV) with the Ll-line of titanium (Ll=395.3 eV). The nitrogen content in the film was calculated from the analytical total assuming that N makes the complement to the analysed elements [55]. Before the EPMA / WDS measurement the cemented carbide substrates were demagnetized.

6.1.2 Glow discharge optical emission spectroscopy

The chemical composition of the coatings was checked by glow discharge optical emission spectroscopy (GDOES) using a SPECTRU-MAT 750 from Leco. GDOES allowed the direct determination of the nitrogen content especially in the Ti-Al-N based coatings.

The concentration depth profile of the coatings was obtained. The measurement included the elements of the coating and the substrate. The Ru content of the coatings could not be measured with SPECTRU-MAT 750 as no analyzer was available.

6.2   X-ray diffraction

6.2.1   Glancing angle X-ray diffraction

Glancing angle X-ray diffraction (GAXRD) was employed to characterise the coatings' microstructure in terms of phase composition, stress-free lattice parameter, macroscopic lattice strain and residual stress, microstrain and crystallite size.

During GAXRD the incident X-ray beam is set to a shallow and fixed angle of incidence γ to the sample surface. The detector moves along the diffractometer circle and records the diffracted intensities from the lattice planes at the appropriate Bragg angle θ. Due to the asymmetric geometry, the angle between the diffracting lattice planes and the sample surface corresponds to ψ and changes permanently with θ (see Eq. (6.1)) [142].

$$\psi = \theta - \gamma \tag{6.1}$$

Due to the change of ψ with half of the diffraction angle 2θ, the residual stress can be determined from the modified $\sin^2 \psi$ plot from one GAXRD pattern. In that case the residual stress is averaged over all measured hkl reflections (see Section 6.2.1.1).

Since in GAXRD the interplanar spacings are measured on different crystallographic planes at different inclinations ψ of the sample, information about the crystal anisotropy of the material can be obtained from one GAXRD pattern (see Section 6.2.1.2).

Another advantage of the GAXRD method is that the penetration depth x_e of the X-rays can be kept constant at different diffraction angles. The penetration depth x_e corresponds to the depth from which the intensity of the diffracted beam is $1/e$ of the incident beam [143] and is determined by the value of the incidence angle γ according to Eq. (6.2):

$$x_e = \frac{\sin \gamma \cdot \sin(2\theta - \gamma)}{\mu(\sin \gamma + \sin(2\theta - \gamma))} \tag{6.2}$$

where μ is the linear absorption coefficient.

The linear absorption coefficient of Cu radiation with a wavelength of 0.15406 nm ($K\alpha_1$) is 492.7 cm^{-1} in a $Ti_{0.5}Al_{0.5}N$ film. The resulting penetration depth of Cu $K\alpha_1$ radiation into a $Ti_{0.5}Al_{0.5}N$ film at an angle of incidence of $\gamma = 3$ ° is about ~ 1 µm and demonstrates the suitability of GAXRD for the investigations of thin films using X-rays.

The GAXRD experiments were performed on two D8 Advanced diffractometers from Bruker AXS. Both diffractometers were equipped with sealed X-ray tubes with copper anode and a parabolic Goebel mirror in the primary beam and with a Soller collimator with an acceptance angle of 0.12 ° and a LiF monochromator in front of a scintillation detector. The Cu Kα_2 / Cu Kα_1 intensity ratio was for one diffractometer 0.08 and for the other 0.3. The angle of incidence of the primary beam at the sample surface was set to either to 2 ° or 3 °.

6.2.1.1 Determination of stress-free lattice parameter and residual stress

In order to determine the stress-free lattice parameter and the residual stress from the GAXRD experiments the modified $\sin^2 \psi$ method was employed for the Ti-Al-N based coatings.

The modified $\sin^2 \psi$ method is based on the "classical" $\sin^2 \psi$ method that considers the lattice spacing d of a certain hkl plane under different inclination angles ψ with respect to the sample surface. The measurement of d of a certain hkl plane as a function of ψ can be realized by the χ-method or Ω-method [144]. The lattice spacing determined by these diffraction experiments is $d_{\phi\psi}^{hkl}$ which is determined from Bragg's law (see Eq. (6.3)):

$$\lambda = 2 d_{\phi\psi}^{hkl} \sin \theta^{hkl} \tag{6.3}$$

where λ is the wavelength, θ^{hkl} is half the diffraction angle and hkl are the Miller indices.

If the sample is strained, then the determined interplanar spacing $d_{\phi\psi}^{hkl}$ deviates from the unstrained lattice spacing d_0. The elastic lattice strain can be described by the comparison of strained and unstrained interplanar spacing according to Eq. (6.4) [145]:

$$\varepsilon_{\varphi\psi} = \frac{d_{\phi\psi} - d_0}{d_0} = (\varepsilon_{11} \cos^2 \phi + \varepsilon_{12} \sin 2\phi + \varepsilon_{22} \sin^2 \phi) \sin^2 \psi$$
$$+ \varepsilon_{33} \cos^2 \psi + (\varepsilon_{13} \cos \phi + \varepsilon_{23} \sin \phi) \sin 2\psi \tag{6.4}$$

where $d_{\phi\psi}$ corresponds to the measured lattice distance at certain angles ϕ and ψ. The directions of the strain components ε_{ij} and the angles ϕ and ψ are visualized in Fig. 6.1.

The strains can be replaced by stresses using Hooke's law according to Eq. (6.5) where the relation between the stress tensor σ_{ij} and the strain tensor ε_{kl} is given by the stiffness tensor C_{ijkl}.

$$\sigma_{ij} = C_{ijkl} \varepsilon_{kl} \tag{6.5}$$

As an alternative, the relation between the stress tensor and the strain tensor can be given by the compliance tensor S_{ijkl} according to Eq. (6.6):

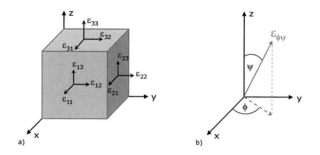

Fig. 6.1: Definition of the strain components (a) with $\varepsilon_{ij} = \varepsilon_{ji}$ and angles ϕ and ψ (b).

$$\varepsilon_{ij} = S_{ijkl}\sigma_{kl} \tag{6.6}$$

The stress tensor and the strain tensor are tensors of the 2nd rank. The stiffness tensor C_{ijkl} and the compliance tensor S_{ijkl} are are tensors of 4th rank with $3^4 = 81$ components. Due to the symmetry of C_{ijkl} and S_{ijkl}, the matrix notation [146] can be used to express Eq. (6.5) and Eq. (6.6) in shorter form with 36 Komponents:

$$\sigma_i = C_{ij}\varepsilon_j \tag{6.7}$$

$$\begin{bmatrix} \sigma_{11} \\ \sigma_{22} \\ \sigma_{33} \\ \sigma_{23} \\ \sigma_{13} \\ \sigma_{12} \end{bmatrix} = \begin{bmatrix} \sigma_1 \\ \sigma_2 \\ \sigma_3 \\ \sigma_4 \\ \sigma_5 \\ \sigma_6 \end{bmatrix} = \begin{bmatrix} C_{11} & C_{12} & C_{13} & C_{14} & C_{15} & C_{16} \\ C_{21} & C_{22} & C_{23} & C_{24} & C_{25} & C_{26} \\ C_{31} & C_{32} & C_{33} & C_{34} & C_{35} & C_{36} \\ C_{41} & C_{42} & C_{43} & C_{44} & C_{45} & C_{46} \\ C_{51} & C_{52} & C_{53} & C_{54} & C_{55} & C_{56} \\ C_{61} & C_{62} & C_{63} & C_{64} & C_{65} & C_{66} \end{bmatrix} \cdot \begin{bmatrix} \varepsilon_1 \\ \varepsilon_2 \\ \varepsilon_3 \\ \varepsilon_4 \\ \varepsilon_5 \\ \varepsilon_6 \end{bmatrix}$$

$$\varepsilon_i = S_{ij}\sigma_j \tag{6.8}$$

$$\begin{bmatrix} \varepsilon_{11} \\ \varepsilon_{22} \\ \varepsilon_{33} \\ 2\varepsilon_{23} \\ 2\varepsilon_{13} \\ 2\varepsilon_{12} \end{bmatrix} = \begin{bmatrix} \varepsilon_1 \\ \varepsilon_2 \\ \varepsilon_3 \\ \varepsilon_4 \\ \varepsilon_5 \\ \varepsilon_6 \end{bmatrix} = \begin{bmatrix} S_{11} & S_{12} & S_{13} & S_{14} & S_{15} & S_{16} \\ S_{21} & S_{22} & S_{23} & S_{24} & S_{25} & S_{26} \\ S_{31} & S_{32} & S_{33} & S_{34} & S_{35} & S_{36} \\ S_{41} & S_{42} & S_{43} & S_{44} & S_{45} & S_{46} \\ S_{51} & S_{52} & S_{53} & S_{54} & S_{55} & S_{56} \\ S_{61} & S_{62} & S_{63} & S_{64} & S_{65} & S_{66} \end{bmatrix} \cdot \begin{bmatrix} \sigma_1 \\ \sigma_2 \\ \sigma_3 \\ \sigma_4 \\ \sigma_5 \\ \sigma_6 \end{bmatrix}$$

Since C_{ij} and S_{ij} are symmetric matrices due to Eq. (6.9) [146], the number of independent stiffness constants and compliances reduces further from 36 to 21.

$$C_{ij} = C_{ji} \quad \text{and} \quad S_{ij} = S_{ji} \tag{6.9}$$

In case of cubic crystals the number of independent stiffness constants and compliances reduces to three and the compliance tensor can be expressed as:

$$[S_{ij}] = \begin{bmatrix} S_{11} & S_{12} & S_{12} & 0 & 0 & 0 \\ S_{12} & S_{11} & S_{12} & 0 & 0 & 0 \\ S_{12} & S_{12} & S_{11} & 0 & 0 & 0 \\ 0 & 0 & 0 & S_{44} & 0 & 0 \\ 0 & 0 & 0 & 0 & S_{44} & 0 \\ 0 & 0 & 0 & 0 & 0 & S_{44} \end{bmatrix} \tag{6.10}$$

In case of elastically isotropic crystallites the macroscopic elastic lattice strain can be rewritten as:

$$\varepsilon_{\phi\psi} = \frac{d_{\phi\psi} - d_0}{d_0} = \frac{1+\nu}{E}(\sigma_{11}\cos^2\phi + \sigma_{12}\sin 2\phi + \sigma_{22}\sin^2\phi - \sigma_{33})\sin^2\psi$$

$$+ \frac{1+\nu}{E}\sigma_{33} - \frac{\nu}{E}(\sigma_{11} + \sigma_{22} + \sigma_{33}) + \frac{1+\nu}{E}(\sigma_{13}\cos\phi + \sigma_{23}\sin\phi)\sin 2\psi \tag{6.11}$$

where E is the Young's modulus and ν is the Poisson's ratio [145]. For elastically isotropic crystallites Eq. (6.11) can be expressed as Eq. (6.14) by using the elastic constants S_1 and $\frac{1}{2}S_2$:

$$S_1 = -\frac{\nu}{E} \tag{6.12}$$

$$\frac{1}{2}S_2 = (1+\nu)/E \tag{6.13}$$

$$\frac{d_{\phi\psi} - d_0}{d_0} = \frac{1}{2}S_2(\sigma_{11}\cos^2\phi + \sigma_{12}\sin 2\phi + \sigma_{22}\sin^2\phi - \sigma_{33})\sin^2\psi$$

$$+ \frac{1}{2}S_2\sigma_{33} + S_1(\sigma_{11} + \sigma_{22} + \sigma_{33}) + \frac{1}{2}S_2(\sigma_{13}\cos\phi + \sigma_{23}\sin\phi)\sin 2\psi \tag{6.14}$$

In case of a quasi-isotropic specimen [144], which is a polycrystal being macroscopically elastically isotropic but it consists of individual crystals being elastically anisotropic, the elastic constants S_1 and $\frac{1}{2}S_2$ in Eq. (6.14) have to be replaced by the X-ray elastic constants s_1^{hkl} and $\frac{1}{2}s_2^{hkl}$ which depend on diffraction indices hkl.

$$\varepsilon_{\phi\psi}^{hkl} = \frac{1}{2}s_2^{hkl}(\sigma_{11}\cos^2\phi + \sigma_{12}\sin 2\phi + \sigma_{22}\sin^2\phi - \sigma_{33})\sin^2\psi$$

$$+ \frac{1}{2}s_2^{hkl}\sigma_{33} + s_1^{hkl}(\sigma_{11} + \sigma_{22} + \sigma_{33}) + \frac{1}{2}s_2^{hkl}(\sigma_{13}\cos\phi + \sigma_{23}\sin\phi)\sin 2\psi \tag{6.15}$$

For macroscopically elastically isotropic specimens with a rotationally symmetric biaxial stress state ($\sigma_{11} = \sigma_{22} = \sigma$) with a zero stress component in the direction normal to the sample surface

($\sigma_{33} = 0$) and the absence of shear stress ($\sigma_{12} = \sigma_{13} = \sigma_{23} = 0$), as it was experimentally proven for CAE TM-Al-(Si)-N coatings (TM = Cr, Ti) [147], Eq. (6.15) can be simplified to:

$$\varepsilon_\psi^{hkl} = \frac{d_\psi^{hkl} - d_0^{hkl}}{d_0^{hkl}} = \frac{1}{2} s_2^{hkl} \sigma \sin^2 \psi + 2 s_1^{hkl} \sigma \qquad (6.16)$$

In general, when the strain ε_ψ^{hkl} is at first calculated from a known d_0^{hkl} and the measured d_ψ^{hkl}, which was obtained at several ψ angles, and then plotted vs. $\sin^2 \psi$, a straight line is obtained. The stress can be derived from the slope of the straight line.

In case of cubic materials the lattice parameter a^{hkl} can be determined from the lattice distance d^{hkl} of the lattice plane with the Miller indices (hkl) using Eq. (6.17):

$$a^{hkl} = d^{hkl} \sqrt{h^2 + k^2 + l^2} \qquad (6.17)$$

Hence, the lattice strain can be expressed by the lattice parameter according to Eq. (6.18)

$$\varepsilon_\psi^{hkl} = \frac{d_\psi^{hkl} - d_0^{hkl}}{d_0^{hkl}} = \frac{a_\psi^{hkl} - a_0}{a_0} \qquad (6.18)$$

The combination of equations (6.16) and (6.18) yields the equation for the $\sin^2 \psi$ plot [143, 148] for cubic materials:

$$a_\psi^{hkl} = a_0 \sigma \left[\frac{1}{2} s_2^{hkl} \sin^2 \psi + 2 s_1^{hkl} \right] + a_0 \qquad (6.19)$$

If the material is isotropic or the XECs are not available then Eq. (6.20) with the bulk elastic properties is used for description of the dependence of a_ψ^{hkl} on $\sin^2 \psi$ [143].

$$a_\psi^{hkl} = a_0 \sigma \left[\frac{1 + \nu}{E} \sin^2 \psi - \frac{2\nu}{E} \right] + a_0 \qquad (6.20)$$

The residual stress can be derived from the slope of the plot a_ψ^{hkl} vs. $\sin^2 \psi$. The lattice parameters $a_\parallel$ and $a_\perp$ lying parallel ($\psi = 90°$) and perpendicular ($\psi = 0°$) to the plane of the film can be obtained from the dependence a_ψ^{hkl} vs. $\sin^2 \psi$ as well. This becomes obvious if Eq. (6.20) is rewritten in the form:

$$a_\psi^{hkl} = a_\perp + (a_\parallel - a_\perp) \sin^2 \psi \qquad (6.21)$$

The stress-free lattice parameter (a_0) [82], or according to Perry [149] the equilibrium lattice parameter, can be obtained from the condition $\sin^2 \psi = 2\nu/(1 + \nu)$ using Eq. (6.22):

$$a_0 = a_\perp + \frac{2\nu}{\nu + 1} (a_\parallel - a_\perp) \qquad (6.22)$$

The macroscopic lattice strain (ε) and the residual stress can be calculated from Eq. (6.23) and Eq. (6.24), respectively [82].

$$\varepsilon = \frac{a_\parallel - a_\perp}{2a_0}$$

(6.23)

$$\sigma = \frac{a_0 - a_\perp}{a_0} \cdot \frac{E}{2\nu}$$

(6.24)

Equations (6.20) to (6.24) were applied to the modified $\sin^2 \psi$ plot that was obtained from the measured GAXRD patterns. In that way the residual stress is averaged over all measured (hkl) reflections.

6.2.1.2 Crystal anisotropy of the lattice deformation

The Cr-Al-(Si)-N coatings and the Ti-Al-N/Al-Ti-Ru-N multilayer coatings were characterized by crystal anisotropy of the lattice deformation of the fcc-phase. This was obvious from a large scatter of the lattice parameter, when the lattice parameter a_ψ^{hkl} that was obtained from GAXRD was plotted vs. $\sin^2 \psi$ [10]. In order to determine the stress-free lattice parameter and the residual stress from the GAXRD pattern in these coatings, the crystal anisotropy had to be considered. This procedure is described in Ref. [137] in detail and is summarized briefly in the following. The XECs s_1^{hkl} and $\frac{1}{2}s_2^{hkl}$ in Eq. (6.19) were substituted by the linear dependence of s_1^{hkl} and $\frac{1}{2}s_2^{hkl}$ on the orientation factor Γ according to Eq. (6.25):

$$s_1^{hkl} = -\frac{\nu^{hkl}}{E^{hkl}} = A_1 + B_1\Gamma$$
$$\frac{1}{2}s_2^{hkl} = \frac{1 + \nu^{hkl}}{E^{hkl}} = A_2 + B_2\Gamma$$

(6.25)

where A_1, A_2, B_1 and B_2 are material constants depending on the single-crystal elastic compliance constants (S_{ij}) of cubic materials and

$$\Gamma = \frac{h^2k^2 + k^2l^2 + l^2h^2}{(h^2+k^2+l^2)^2}$$

(6.26)

is the cubic invariant. Hence, the measured lattice parameter a_ψ^{hkl} depends on $\sin^2 \psi$ and Γ:

$$a_\psi^{hkl} = A_2 a_0 \sigma \sin^2 \psi + B_2 a_0 \sigma \Gamma \sin^2 \psi + 2B_1 a_0 \sigma \Gamma + 2A_1 a_0 \sigma + a_0 \qquad \textbf{(6.27)}$$

The experimental data (a_ψ^{hkl}) were fitted by Eq. (6.27) that yielded four parameters P_1 to P_4:

$$P_1 = A_2 a_0 \sigma$$

(6.28)

$$P_2 = B_2 a_0 \sigma$$

(6.29)

$$P_3 = 2B_1 a_0 \sigma$$

(6.30)

$$P_4 = a_0(2A_1\sigma + 1) \tag{6.31}$$

The parameters P_1 to P_4 were used to correct the crystal anisotropy of the lattice parameters. This was done by the subtraction of the Γ-dependent part from the measured lattice parameters a_ψ^{hkl}. The Γ-independent part can be described by Eq. (6.32) which yields the lattice parameters a_ψ^{h00} that correspond to the $(h00)$ planes since the orientation factor Γ equals zero for the lattice planes $(h00)$.

$$a_\psi^{h00} = A_2 a_0 \sigma \sin^2\psi + 2A_1 a_0 \sigma + a_0 \tag{6.32}$$

The dependence a_ψ^{h00} vs. $\sin^2\psi$ is now a linear function.

Taking the Reuss model (see e.g. References [144, 150]) into account, the XECs can be calculated from the single-crystal compliance tensor S_{ij} (see Eq. (6.8)):

$$s_1^{hkl} = S_{12} + S_0\Gamma$$
$$\frac{1}{2}s_2^{hkl} = S_{11} - S_{12} - 3S_0\Gamma \tag{6.33}$$

where

$$S_0 = S_{11} - S_{12} - \frac{1}{2}S_{44} \tag{6.34}$$

Comparing Equations (6.33) and (6.25), the coefficients A_1, A_2, B_1 and B_2 can be related to the single elastic compliance constants S_{11}, S_{12} and S_{44} of cubic materials (see Eq. (6.10)) as follows:

$$A_1 = S_{12} \tag{6.35}$$

$$A_2 = S_{11} - S_{12} \tag{6.36}$$

$$B_1 = S_{11} - S_{12} - \frac{1}{2}S_{44} = S_0 \tag{6.37}$$

$$B_2 = -3\left(S_{11} - S_{12} - \frac{1}{2}S_{44}\right) = -3S_0 \tag{6.38}$$

Within the Reuss approximation Eq. (6.27) can be expressed as Eq. (6.39) and the Γ-independent part of the function of the lattice parameter on $\sin^2\psi$ as shown in Eq. (6.32) can be written as Eq. (6.40):

$$a_\psi^{hkl} = a_0\sigma(S_{11} - S_{12})\sin^2\psi - 3a_0\sigma S_0\Gamma\sin^2\psi$$
$$+ 2a_0\sigma S_0\Gamma + 2a_0\sigma S_{12} + a_0 \tag{6.39}$$

$$a_\psi^{h00} = a_0 \sigma \underbrace{(S_{11} - S_{12})}_{\frac{1}{2}s_2^{100}} \sin^2 \psi + 2a_0 \sigma \underbrace{S_{12}}_{s_1^{100}} + a_0 \tag{6.40}$$

Consequently, a_0 and σ can be determined from the linear dependence of a_ψ^{h00} vs. $\sin^2 \psi$ if the XECs s_1^{100} and $\frac{1}{2}s_2^{100}$ are known.

Furthermore, Eq. (6.39) can be used for the calculation of the degree of the anisotropy (A) of the elastic constants of the cubic phase that is defined by Eq. (6.41):

$$A = \frac{2C_{44}}{C_{11} - C_{12}} = \frac{2(S_{11} - S_{12})}{S_{44}} = \frac{(S_{11} - S_{12})}{(S_{11} - S_{12} - S_0)} \tag{6.41}$$

where C_{11}, C_{12} and C_{44} are the stiffnesses of a single crystal.

In Eq. (6.41) $A = 1$ means no anisotropy. If $A < 1$, the elementary cell can be deformed more easily in the $\langle hhh \rangle$ directions than in the $\langle h00 \rangle$ directions. In case of compressive residual stress and $A < 1$, the lattice parameter a^{111} is larger than a^{200}. If $A > 1$, the elementary cell can be deformed more easily in the $\langle h00 \rangle$ directions than in the $\langle hhh \rangle$ directions and a^{200} is larger than a^{111} in case of compressive residual stress.

The anisotropy factor (A) can be directly calculated from the fitting parameters P_1 and P_3 according to Eq. (6.42) after combining Eq. (6.28) and Eq. (6.36) as well as Eq. (6.30) and Eq. (6.37):

$$A = \left[1 - \frac{S_0}{S_{11} - S_{12}}\right]^{-1} = \left[1 - \frac{P_3}{2 \cdot P_1}\right]^{-1} = \left[1 - \frac{2a_0 \sigma S_0}{2a_0 \sigma (S_{11} - S_{12})}\right]^{-1} \tag{6.42}$$

In addition, the ratio between the fitting parameters P_2 (see Eq. (6.29)) and P_3 (see Eq. (6.30)) can be used as an indicator of the departure of the real elastic behaviour from the Reuss model, since Eq. (6.27) is interrelated with Eq. (6.39) by the Reuss approximation. In case of elastic behaviour according to the Reuss model, the ratio between P_2 and P_3 (see Eq. (6.43)) would correspond to $-3/2$.

$$\frac{P_2}{P_3} = \frac{B_2 a_0 \sigma}{2 B_1 a_0 \sigma} = \frac{-3a_0 \sigma S_0}{2a_0 \sigma S_0} = \frac{-3a_0 \sigma \left(S_{11} - S_{12} - \frac{1}{2} S_{44}\right)}{2a_0 \sigma \left(S_{11} - S_{12} - \frac{1}{2} S_{44}\right)} \tag{6.43}$$

6.2.2   *In situ* synchrotron HT-GAXRD

In situ synchrotron high temperature (HT) GAXRD experiments were performed at the Rossendorf Beamline at the European Synchrotron Radiation Facility in Grenoble. In the frame of this work two sample series were investigated using *in situ* synchrotron HT GAXRD experiments. One sample series were Ti-Al-N monolayer coatings and the other sample series were Ti-Al-N / Al-Ti-Ru-N multilayer coatings. The experimental set up for the two sample series differed slightly with respect to the wavelength, irradiated area, incidence angle, detector, type of the BN heater and applied time-temperature profile. For both sample series the HT-GAXRD experiments were done at a pressure of $\sim 10^{-3}$ Pa in a high-temperature chamber that was covered with a hemispherical beryllium dome. The high-temperature chamber was equipped with a heated BN sample holder. Prior annealing, the coated SNUN type cutting inserts were cut to 500 to 600 µm thick plates in order to obtain a quick and homogeneous heating of the coating. The plates were placed with the substrate side on the BN sample holder. The temperature was checked by a thermocouple that was inserted in the BN heater plate. Due to the experimental set up, the movement of the detector arm was restricted until a maximum of 85° 2θ.

In case of the Ti-Al-N monolayer coatings, the wavelength was 0.08857 nm and the incidence angle of the primary beam on the sample surface was set to 0.5°. The diffracted intensity was counted by a scintillation detector in the 2θ-range from 17° to 70°. The applied time-temperature profile is shown schematically in Fig. 6.2a. Before annealing the coatings were measured in the as-deposited state at 20°C. The HT-GAXRD experiments were done at 450°C, 650°C and 850°C consecutively. After each annealing step the samples were cooled to 100°C to eliminate the effect of thermal expansion on the stress-free lattice parameter. The HT-GAXRD experiments at 850°C were repeated three times in order to get insight into the time dependent microstructure changes. The time scale of the experiments were determined by ~ 30 min heating to the requested temperature, ~ 40 min for each GAXRD measurement and ~ 30 min cooling from the elevated temperature to ~ 100°C.

In case of the Ti-Al-N / Al-Ti-Ru-N multilayer coatings, the wavelength for HT-GAXRD experiments was 0.107814 nm. The incidence angle of the primary beam on the sample surface was set to 2°. The height and the width of the incident beam were limited to 300 µm. The diffracted intensity was counted by a one-dimensional position sensitive detector in the 2θ-range from 17° to 83°. The time-temperature profile used for *in situ* HT-GAXRD measurements of the Ti-Al-N / Al-Ti-Ru-N multilayer coatings is shown in Fig. 6.2b. At first the coatings were measured in the as-deposited state at 20°C. Then the samples were annealed at 650°C,

850°C, 900°C and 950°C for 60 min. After a dwell time of 15 min at the respective temperature a short GAXRD scan (17° to 47°) and two complete GAXRD measurements (17° to 83°) were done consecutively. After each annealing step the samples were cooled to 100°C to eliminate the effect of thermal expansion on the stress-free lattice parameter.

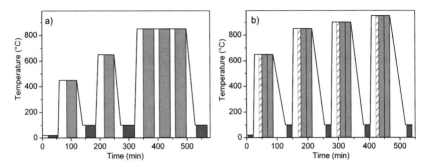

Fig. 6.2: Schematic representation of the applied temperature profile during the *in situ* synchrotron HT-GAXRD experiments of the Ti-Al-N monolayer coatings (a) and Ti-Al-N / Al-Ti-Ru-N multilayer coatings (b). The colored columns correspond to complete GAXRD measurements. The striped columns denote short GAXRD measurements. The white regions correspond to the adjustment, heating, cooling and dwell time.

6.2.3 Pole figure measurements

The preferred orientation of the fcc-crystallites in the coatings was determined by measuring the pole figures $\{111\}_{fcc}$, $\{100\}_{fcc}$ and $\{110\}_{fcc}$ of the fcc-phase on a PTS 3000 X-ray diffractometer from Seifert equipped with an Eulerian cradle and with a mono-capillary in the primary beam. For the pole figure measurement the samples were tilted by the polar angle ψ from 0° to 75° with a step size of 5°. For each inclination the samples were rotated around the surface normal by the azimuth angle ϕ from 0° to 360° with a step size of 5° and the intensity of the respective 2θ position was measured. A background correction was done for every inclination. The measured pole figures were re-calculated to the stereographic projection according to Eq. (6.44) and Eq. (6.45) [151]:

$$x = \frac{\sin \psi}{1 + \cos \psi} \cdot \cos \phi \qquad\qquad (6.44)$$

$$y = \frac{\sin \psi}{1 + \cos \psi} \cdot \sin \phi \qquad\qquad (6.45)$$

The inclination of the $\langle 100 \rangle_{fcc}$ direction from the surface normal was determined from the intensity profile of the $\{100\}_{fcc}$ pole figure along a rotation angle ϕ_1 and $\phi_1 + 180°$ over all inclination angles ψ. The intensity profile was fitted by two Gaussian functions. The equation of the Gaussian function is given in Eq. (6.46) where the four parameters y_0, x_c, A and w have the following meaning. The parameter y_0 is the base of the curve, x_c is the center, A is the area and w is the full width and half maximum (FWHM). Thus, the inclination angle of the $\langle 100 \rangle_{fcc}$ direction corresponds to x_c.

$$y = y_0 + \left(\frac{A \cdot \exp(-2((x - x_c)/w)^2)}{w \cdot \sqrt{\pi/2}} \right) \tag{6.46}$$

The angle between the lattice plane $(h_1 k_1 l_1)$ and $(h_2 k_2 l_2)$ in the cubic system was calculated according to Eq. (6.47). Then the calculated angles between the $\{111\}_{fcc}$, $\{100\}_{fcc}$ and $\{110\}_{fcc}$ planes were compared with the measured polar angles.

$$\phi = \arccos \left(\frac{h_1 h_2 + k_1 k_2 + l_1 l_2}{\sqrt{h_1^2 + k_1^2 + l_1^2} \cdot \sqrt{h_2^2 + k_2^2 + l_2^2}} \right) \tag{6.47}$$

6.2.4   Reciprocal space maps and XRD rocking curves

Reciprocal space maps (RSMs) were recorded for the TiN / AlN / TiN layer stacks grown on $(1\bar{1}0)_{fcc}$ and $(00\bar{1})_{fcc}$ oriented MgO substrates (for details see Chapter 7.3.1) in order to visualize the areas around reciprocal lattice points in reciprocal space.

Usually a high resolution X-ray diffractometer is used for the measurement of RSMs. A triple-axis diffractometer from Seifert FPM which is equipped with an Eulerian cradle, an sealed X-ray tube with copper anode, a (111) oriented Si monochromator crystal in the primary beam and a (111) oriented Si analyzer crystal in the diffracted beam was available within this work. The cross section of the primary X-ray beam was reduced by slits to a size of 0.09×4 mm^2. The wavelength of the used Cu Kα_1 radiation corresponds to $\lambda_{\alpha 1} = 0.15406$ nm. In order to increase the detected intensity of the diffracted beam, the Si analyzer was removed from the beam path. Still, the diffracted intensity was not sufficient for the measurement of all desired layers (MgO, TiN or AlN) and lattice planes. Thus, just the RSMs of the $(1\bar{1}0)_{fcc}$ planes of the MgO substrate and the fcc-TiN layer deposited onto the $(1\bar{1}0)_{fcc}$ oriented MgO substrate as well as the $(00\bar{1})_{fcc}$ planes of the MgO substrate and the fcc-TiN layer deposited onto the $(00\bar{1})_{fcc}$ oriented MgO substrate could be recorded using the high resolution diffractometer. High resolution XRD was also used to measure the XRD rocking curves for the $(01\bar{3})_w$ lattice planes of the w-AlN layer included in the TiN / AlN / TiN layer stacks, which was deposited onto the $(1\bar{1}0)_{fcc}$ oriented MgO substrates, respectively.

Due to the low intensity of the diffracted X-rays in the HRXRD experiment, all desired RSMs and rocking curves were measured with a D8 ADVANCE diffractometer from Bruker AXS. The diffractometer is equipped with a sealed X-ray tube with copper anode, a parabolic Goebel mirror in the primary beam and with a Soller collimator with an acceptance angle of 0.12° and a LiF monochromator. The Cu Kα2 / Cu Kα1 intensity ratio was 0.15 and the wavelength of the Cu Kα1 radiation was $\lambda_{\alpha 1}$. = 0.15406 nm.

Since the sample holder of the D8 ADVANCE diffractometer could not be rotated around the ϕ axis, the edges of rectangular shaped MgO single crystal were used for the manual alignment of the sample in the sample holder. The edges of the $(1\bar{1}0)_{fcc}$ oriented MgO crystal were aligned parallel to the $(001)_{fcc}$ and $(110)_{fcc}$ planes (see Fig. 7.42a). The edges of the $(00\bar{1})_{fcc}$ oriented MgO crystal were aligned parallel to the $(010)_{fcc}$ and $(100)_{fcc}$ planes (see Fig. 7.42b). Additionally, each sample position was checked by a symmetrical scan of the MgO substrate and two 2θ scans of an $\{hkl\}_{fcc}$ plane of the MgO sybstrate that is inclined by $+\psi$ and $-\psi$ from the sample surface and whose normale is lying in the same plane like the wavevector of the incident and diffracted beam.

The RSMs and XRD rocking curves obtained from the D8 ADVANCE diffractometer were used to compare the desorientation of the $(01\bar{3})_w$ and $(102)_w$ lattice planes of the w-AlN layer with respect to the TiN / AlN interface.

During the reciprocal space mapping a set of XRD rocking curves for different values of 2θ were measured. The step sizes $\Delta 2\theta$ and $\Delta\omega$ were 0.02° for the diffraction lines of the MgO substrate and TiN layer and 0.05° for the diffraction lines of the AlN layer.

The obtained data set of the real space $(\omega, 2\theta)$ were converted to the q_x, q_z coordinates of the scattering vector $\vec{q}$ in reciprocal space according to the Eqs. (6.48) and (6.49) [152]:

$$q_x = \frac{2\pi}{\lambda}(\cos(2\theta - \omega) - \cos(\omega)) \qquad \textbf{(6.48)}$$

$$q_z = \frac{2\pi}{\lambda}(\sin(\omega) + \sin(2\theta - \omega)) \qquad \textbf{(6.49)}$$

where λ corresponds to the used wavelength, ω is the angle between the incident beam and the sample surface and 2θ is the diffraction angle.

6.3   Transmission electron microscopy

Investigations using transmission electron microscopy (TEM) were done on two analytical high resolution transmission electron microscopes from Jeol: JEM 2010 FEF and JEM 2200 FS. Both microscopes are equipped with a field emission gun that operates with an acceleration voltage of 200 kV, as well as an ultra-high-resolution objective lens with a spherical aberration of $C_s = 0.5$ mm and an in-column energy filter, called omega filter. In contrast to the JEM2010 FEF, the JEM 2200 FS is equipped with a C_s corrector which corrects the spherical aberration of the condenser lense. This correction offers an increased point resolution of 0.19 nm.

The Cr-Al-Si-N and Ti-Al-N monolayer coatings were investigated in plane view orientation using JEM 2010 FEF. The Ti-Al-N / Al-Ti-(Ru)-N multilayer coatings were prepared in cross-section by focused ion beam technique and were studied using the JEM 2200 FS.

Basics and details about TEM are given e.g. in References [153-157]. Therefore, the TEM methods, which were used within this work, are explained very briefly in the following.

6.3.1   Imaging using TEM

Imaging of the samples was done using TEM, scanning transmission electron microscopy (STEM) and high resolution transmission electron microscopy (HRTEM).

The images in TEM and STEM are determined by two principle contrasts: the mass-thickness contrast and the diffraction contrast [155].

The **mass-thickness contrast** results from incoherent elastic scattering of electrons. An increasing atomic number Z leads to an increasing cross section for Rutherford scattering. On the other hand more elastic scattering events take place in a thicker TEM foil. Thus high Z-regions scatter more than low Z-regions and thicker regions scatter more electrons than thinner areas [155]. The strongly scattered electrons are removed by the objective aperture from the beam path in case of bright field (BF) imaging. Hence, only the weakly scattered electrons, which move nearly parallel to the optical axis, contribute to the image formation due to the central position of the objective aperture around the direct beam. In case of a dark field (DF) imaging, the strongly scattered electrons are selected for image formation. In TEM mode, the objective aperture is used to select the strongly scattered electrons.

In case of a crystalline sample, scattering can also occur if incoming electrons hit the lattice planes at the Bragg angle. The crystallites that fulfil the Bragg condition appear darker in case of a BF image and form the **diffraction contrast**. A special case of the diffraction contrast is the so called two-beam condition. In that case the specimen is tilted until only one diffracted beam is strong in the diffraction pattern; the second beam is the direct beam. Consequently, only the area with a specific set of *hkl* being at the Bragg condition appears bright in the corresponding DF image.

STEM imaging was used mainly in DF mode in order to visualize changes in the Al concentration across the multilayer stacks in Ti-Al-N / Al-Ti-(Ru)-N multilayers. The C_s corrected JEM 2200 FS was used for the investigations in STEM mode which allowed to focus the electron beam to diameter of ~1 nm. The convergent electron beam is than scanned across the specimen. The DF STEM images were recorded with a high-angle annular dark field (HAADF) detector. The HAADF detector collects only electrons that are scattered in an angle range of 100 to 170 mrad from the optical axis [158]. Hence, the portion of the electrons that are scattered by the Bragg condition (usually a few mrad) and which hit the HAADF detector is low. Consequently, the contrast in a HAADF STEM image is mainly determined by the mass of the elements.

HRTEM images were taken in order to determine the local orientation of phases (see Chapter 6.3.2). The contrasts seen in a HRTEM image are a result of the phase shift of the electron wave, which is caused by the interaction of electron wave with the positive potential of the crystal. If the phase modulated waves interfere with the scattered waves, the amplitudes change. The amplitude modulation can then be registered by a CCD camera [157]. The size of the phase shift determines whether the atom columns or the space between them appear bright in the HRTEM image.

6.3.2 Determination of the local orientation

Fast Fourier transformation (FFT) was applied to HRTEM images. In that way periodic structures present in the HRTEM image could be transformed to a quasi diffraction pattern. The FFT patterns were used to investigate the local orientation of the phases and their interfaces. Orientation determination of sample regions with a diameter of at least ~ 100 nm was done by selected area electron diffraction (SAED).

The basics of electron diffraction in TEM and FFT are described in detail in [154] and [155], respectively. Therefore, the procedure of the orientation determination of the observed phases in the TEM foil from the SAED and FFT patterns is described very briefly in the following with the help of Fig. 6.3, which shows schematically a point diagram from a single crystal that was observed by SAED or FFT of a HRTEM image.

The diffraction spots or reflections seen in the SAED or FFT patterns correspond to reciprocal lattice points. The origin (000) of the reciprocal lattice is labelled as O in Fig. 6.3. The vector $\vec{g}_{hkl}$ from the origin (000) to a diffraction spot is described by:

$$\vec{g}_{hkl} = h\vec{a}^* + k\vec{b}^* + l\vec{c}^* \tag{6.50}$$

where $\vec{a}^*$, $\vec{b}^*$ and $\vec{c}^*$ are the unit-cell translation vectors in reciprocal space. The reciprocal lattice vector $\vec{g}_{hkl}$ is perpendicular to the plane with the Miller indices (hkl) [159] and its length $|\vec{g}_{hkl}|$ corresponds to the inverse lattice spacing d_{hkl}^{-1} of that lattice plane (see Eq. (6.51)).

$$|\vec{g}_{hkl}| = d_{hkl}^{-1} \tag{6.51}$$

Furthermore, the angle φ between two vectors $\vec{g}_1$ and $\vec{g}_2$ in the reciprocal space equals the angle φ between the lattice planes $(h_1 k_1 l_1)$ and $(h_2 k_2 l_2)$ in real space and can be calculated from Eq. (6.52):

$$\cos \varphi = \frac{\vec{g}_1 \cdot \vec{g}_2}{|\vec{g}_1| \cdot |\vec{g}_2|} \tag{6.52}$$

From the diffraction patterns obtained by SAED or FFT, the distances $|\vec{g}_i|$ of i-reflections ($i = 1, 2, \ldots, n$) from the origin (000) were determined as well as the angle between the vectors $\vec{g}_i$. This was done using $DigitalMicrograph^{TM}$ from Gatan [160]. Since the measured angles and inverse lattice spacings are characteristic for the fcc-(Ti,Al)N, fcc-(Cr,Al)N and w-AlN phase, the observed diffraction pattern could be assigned to the respective phase. This was done by the comparison of the observed diffraction pattern with the simulated one which was obtained either by the program $SingleCrystal^{TM}$ [161] or by the program $JEMS$ [162].

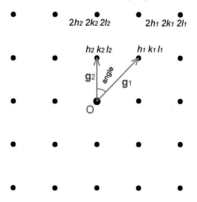

Fig. 6.3: Schematic drawing of a single diffraction pattern obtained by SAED or FFT showing the vectors $\vec{g}_1$ and $\vec{g}_2$ and the angle φ between them. The letter O refers to the direct beam and corresponds to the (000) reflection.

Once two vectors $\vec{g}_1$ and $\vec{g}_2$ of a single diffraction pattern of one phase are determined (c.f. Fig. 6.3) the direction of the incident beam $\vec{B}_{ebeam}$ can be calculated from their cross product. In determinant form the cross product of $\vec{g}_1$ and $\vec{g}_2$ can be expressed by Eq. (6.53) [159, 163]. The order of the vectors in Eq. (6.53) should be chosen in that way, that if a circle is drawn around the origin (000) then the vectors $\vec{g}$ should be numbered counterclockwise [154].

$$\vec{g}_1 \times \vec{g}_2 = \frac{1}{V}\begin{vmatrix} \mathbf{a} & \mathbf{b} & \mathbf{c} \\ h_1 & k_1 & l_1 \\ h_2 & k_2 & l_2 \end{vmatrix} = V^* \begin{vmatrix} \mathbf{a} & \mathbf{b} & \mathbf{c} \\ h_1 & k_1 & l_1 \\ h_2 & k_2 & l_2 \end{vmatrix}$$

$$= V^* \left((h_1 l_2 - k_2 l_1)\vec{a} + (l_1 h_2 - l_2 h_1)\vec{b} + (h_1 k_2 - h_2 k_1)\vec{c} \right) = V^* \vec{B}_{ebeam}$$

(6.53)

The symbol V denotes the volume of the unit cell in real space and V^* corresponds to the volume of the reciprocal unit cell. The bold written symbols $\mathbf{a}$, $\mathbf{b}$ and $\mathbf{c}$ represent the direct basis vectors in real space. Equation (6.53) is valid for all crystal systems [159]. Since for the determination of the direction of the incident beam only the direction of the vector is important and not its actual length, the volume of the unit cell or the reciprocal volume can be dropped and the calculation of the incident beam direction can be reduced to Eq. (6.54).

$$\vec{B}_{ebeam} = \begin{bmatrix} k_1 l_2 - k_2 l_1 \\ l_1 h_2 - l_2 h_1 \\ h_1 k_2 - h_2 k_1 \end{bmatrix}$$

(6.54)

By convention the vector $\vec{B}_{ebeam}$ is antiparallel to the travelling direction of the electron and points up the column of the TEM [154]. Furthermore, the vector $\vec{B}_{ebeam}$ is parallel to the TEM foil normal and corresponds to the zone axis. The zone axis is a direction $[uvw]$ which is common to all planes of the zone. A plane (hkl) belongs to a zone with the zone axis $\vec{t} = [u, v, w]$ if the condition given in Eq. (6.55) is fulfilled. This means that a plane belongs to the $[uvw]$ zone if the normal of the plane (which is equivalent to $\vec{g}$ in Eq. (6.55)) is perpendicular to the zone axis $[uvw]$ (which is equivalent to $\vec{t}$ in Eq. (6.55)). Equation (6.55), which is the so called "Weiss law", is valid for all crystal systems [159].

$$\vec{g} \cdot \vec{t} = 0$$

$$\begin{bmatrix} ha^* \\ kb^* \\ lc^* \end{bmatrix} \cdot \begin{bmatrix} ua \\ vb \\ wc \end{bmatrix} = hu + kv + lw = 0$$

(6.55)

6.3.3   Analytical TEM

In order to investigate local concentration fluctuations across a multilayer stack of the Ti-Al-N / Al-Ti-(Ru)-N multilayer coatings, energy dispersive spectroscopy (EDS) and electron energy loss spectroscopy (EELS) were done in STEM mode with the Cs corrected JEM 2200 FS offering an electron probe with a diameter of 0.5 nm.

6.3.3.1   Energy dispersive spectroscopy

EDS was done along a line scan across several multilayer stacks in the Ti-Al-N / Al-Ti-(Ru)-N multilayer coatings with a step width in the range of 1 nm. The acquisition time for EDS was 20 to 30 s of each measuring point. In order to correct the sample drift, a drift correction was applied usually after each third measuring point. In most cases EDS and EELS were performed at once during the measurement along a line scan.

From the EDS measurements the atomic ratios [Me] / ([Ti] + [Al] + [Ru]) of the metallic components (Me = Ti, Al or Ru) were determined. For that purpose, the X-ray characteristic peaks of the Al Kα line at 1.49 keV, the Ti Kα line at 4.51 keV and the Ru Kα line at 19.23 keV were chosen. The analyses of the measured energy dispersive spectra were done with the EDS tool implemented within the *DigitalMicrograph*^{TM} software from Gatan [160]. The quantification mode was standardless and applied the correction model from Cliff and Lorimer. According to the Cliff - Lorimer model, the intensity ratio of the measured characteristic X-ray peaks I_A and I_B of the elements A and B is proportional to the ratio of their atomic concentration c_A and c_B, see Eq. (6.56). The proportional coefficient k_{AB} is the Cliff - Lorimer k-factor [157].

$$\frac{I_A}{I_B} = k_{AB} \cdot \frac{c_A}{c_B} \qquad \qquad \textbf{(6.56)}$$

Theoretical k factors [164] were used in the standardless quantification performed with the *DigitalMicrograph*^{TM} software from Gatan. The approach given in Eq. (6.56) can be extended to several elements, see e.g. [157].

Among the available correction models within the software package, the Cliff - Lorimer model yielded the most reliable atomic ratios concerning the coating composition. Nevertheless, the [Al] / ([Ti] + [Al] + [Ru]) ratio averaged over multiple bi-layer periods yielded often a lower value as expected from the average Al-ratio determined by EPMA / WDS and GDOES. However, the quantification of the metallic ratios from the EDS signals across the multilayer stacks was done mainly to reveal differences in the composition along a line scan rather than to determine the absolute composition. Furthermore, it has to be kept in mind that indeed the diameter of the electron probe was small (~ 0.5 nm), but the spatial resolution is lower. This is attributed to the broadening of the electron beam due to elastic scattering as it passes through the TEM foil. The mean broadening of the electron beam b_{ebeam} given in nm can be estimated after Goldstein [165] according to Eq. (6.57).

$$b_{ebeam} = 625 \cdot 10^7 \cdot {}^Z\!/_{E_0} \cdot \sqrt{{}^\rho\!/_{A_{aw}} \cdot t^3} \qquad (6.57)$$

where Z is the mean atomic number, E_0 is the incident electron energy in keV, ρ is the density in g/cm^3, A_{aw} is the atomic weight and t is the TEM foil thickness in cm. The influence of the TEM foil thickness and the density of the investigated material on the broadening of the electron beam with an energy of 200 keV is shown in Fig. 6.4.

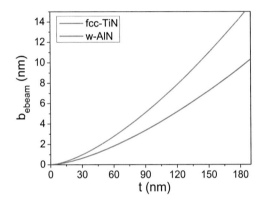

Fig. 6.4: Broadening of the electron beam (b_{ebeam}) as function of the TEM foil thickness (t) in a fcc-TiN and w-AlN specimen as estimated using Eq. (6.57). The energy of the primary electrons is 200 keV.

Since the mean atomic number, atomic weight and density can range between those of fcc-TiN ($Z = (22 + 7)/2 = 14.5$, $\rho =5.393$ g/cm^3, $A_{aw} = (47.9 + 14)/2 \approx 31$) and w-AlN ($Z = (13 + 7)/2 = 10$, $\rho = 3.262$ g/cm^3, $A_{aw} = (27 + 14)/2 = 20.5$) in the investigated Ti-Al-N / Al-Ti-N multilayers, these two phases were chosen to visualize the upper and lower limit of the electron beam broadening in Fig. 6.4. The estimated broadening is ~ 6 nm and ~ 4 nm in case of a 100 nm thick TEM foil consisting of fcc-TiN and w-AlN, respectively.

6.3.3.2 Electron energy loss spectroscopy

Electron energy loss (EEL) spectra were recorded around the nitrogen K edge (~ 400 eV) as well as in the low loss region including the zero loss (ZL) peak. The ZL corresponds to the electrons that travelled through the microscope without any energy transfer to the TEM foil. Both spectra were recorded in order to obtain information about the nitrogen concentration, the crystal structure especially of AlN [166, 167] and the TEM foil thickness, respectively. The recording and the analysis of both spectra are explained in the following.

Recording of the EEL spectra in the range of the nitrogen K edge

The EEL spectra were recorded with three alternative dispersions of the Ω filter using a dispersion of 80 μm/eV, 100 μm/eV or 200 μm/eV. An example of two raw EEL spectra measured in fcc-TiN with a dispersion of 100 μm/eV and 200 μm/eV are shown in Fig. 6.5a. With increasing dispersion from 80 μm/eV to 200 μm/eV the energy range covered by the CCD decreased. The energy range covered by the CCD was ~ 240 eV, ~ 220 eV and ~ 110 eV for the dispersion of 80 μm/eV, 100 μm/eV and 200 μm/eV, respectively. In order to record simultaneously the core loss region of the nitrogen K edge near 400 eV (depending on the phase) [168] and of the titanium $L_{2,3}$ edge at 455.5 eV [168], the energy dispersive plane was shifted by the Ω filter usually in the range of 430 eV to 500 eV.

The EEL spectra at the the nitrogen K edge and the titanium $L_{2,3}$ edge were measured using condenser lense CL3 yielding a convergence semi-angle α_{conv} of 22 mrad [169] (see Fig. 6.5b).

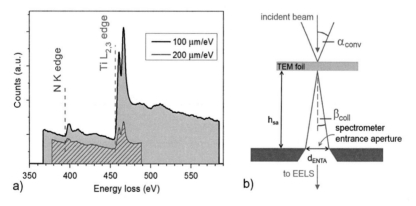

Fig. 6.5: Comparison of EEL spectra of a fcc-TiN layer (sample FIB_2) measured with a dispersion of 100 μm/eV and 200 μm/eV (a) showing the nitrogen K edge and the titanium $L_{2,3}$ edge. The energy dispersive plane was shifted by 500 eV and 460 eV, respectively. Schematic presentation of the convergence semi angle α_{conv} and the collection semi-angle in β_{coll} in a STEM device (b), reproduced after [156].

The spectrometer entrance aperture ENTA4 was used in order to improve the spatial resolution of the EELS signal. Its diameter (d_{ENTA}) and the distance from the specimen to the aperture (h_{sa}) determined the collection semi-angle β_{coll} of the incoming beam which was 5.5 mrad [169].

The measurement of the EEL spectra was done along a line scan across several multilayer stacks in the Ti-Al-N / Al-Ti-(Ru)-N multilayer coatings. The step width was in the range of 1 nm. The acquisition time for EELS was 20 to 30 s of each measuring point. As in the case of the EDS line scans, a drift correction was done as well.

Recording of the EEL spectra in the low loss region including the ZL peak

Parallel to the EELS line scan that recorded the electron energy loss around the nitrogen K edge and the titanium $L_{2,3}$ edge, an EELS line scan that recorded the zero loss (ZL) and the low loss region was done. Depending on the used dispersion, the measured energy range expanded from 0 eV to ~ 110 eV (for a dispersion of 200 µm/eV) or from 0 eV to ~ 220 eV (for a dispersion of 100 µm/eV) or from 0 eV to ~ 240 eV (for a dispersion of 80 µm/eV). In order to operate the CCD safely, the spectra with the ZL peak were measured using the high contrast aperture HCA4 which decreased the number of electrons arriving at the CCD. Furthermore, the acquisition time for the recording of the spectra was adjusted with respect to the maximum intensity of the ZL peak which should not exceed 30000 counts per pixel. Due to that the measuring time for each spectrum was usually in the range of 0.1 to 0.5 s. The used HCA4 yielded a collection semi-angle β_{coll} of 1.8 mrad [169]. The convergence semi-angle α_{conv} was 22 mrad [169] (see Fig. 6.5b).

Calibration of the EEL spectra

The energy resolution in the EEL spectra, which is influenced by the spectrometer and the energy distribution of the primary electrons, could be deduced from the FWHM of the ZL peak. It ranged between 0.8 to 1.2 eV.

The energy resolution per pixel of the CCD varied for different TEM sessions. In order to consider this effect, a calibration procedure was done every TEM session by recording the titanium $L_{2,3}$ edge or the ZL peak with different energy shifts. The measured intensity of each spectrum was plotted versus the number of the CCD camera pixel. Then all spectra were summated as shown in Fig. 6.6. In the example shown in Fig. 6.6, each single spectrum was recorded with a dispersion of 200 µm/eV and the energy was shifted from 0 to 100 eV with an increment of 10 eV (E_{iES}). Afterwards the difference of the number of the pixel (Δpix_i)

between the maximum of the ZL peaks of adjacent EEL spectra was determined. Then the calibration factor (f_{EELS}) was calculated according to Eq. (6.58).

$$f_{EELS} = \frac{\sum_{i=1}^{j} {}^{E_{iES}}/_{\Delta\, \text{pix}_i}}{j} \qquad (6.58)$$

The calibration procedure using the titanium $L_{2,3}$ edge works analogous. The titanium $L_{2,3}$ edge was recorded while shifting the energy dispersive plane from 440 to 540 eV with the Ω filter with an increment of 10 eV (E_{iES}) for a dispersion of 200 μm/eV.

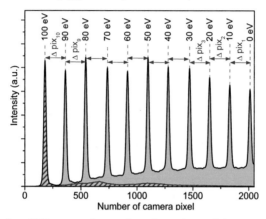

Fig. 6.6: Sum of ten EEL spectra (cyan coloured curve) containing the ZL peak that were recorded for different energy shifts in the range of 0 eV to 100 eV (see label above) as function of the number of the CCD camera pixel. The single spectrum obtained for an energy shift of 100 eV is shown as striped curve.

After obtaining the calibration factor f_{EELS}, the energy scale of the measured EEL spectra containing the nitrogen K edge and the titanium $L_{2,3}$ edge that were intended for further analysis could be calibrated according to the expected position of the titanium $L_{2,3}$ edge (455.5 eV) and the calibration factor f_{EELS}. The energy scale of the measured EEL spectra containing the ZL peak were calibrated according to the expected position of the ZL peak (0 eV) and f_{EELS}.

Extraction of the EEL signal of the N K edge from the measured spectra

For the analysis of the N K edge, an EELS line scan recording the EEL around the N K edge (core edge) was done at first. Additionally, an EELS line scan, which recorded the ZL signal and the low loss region, was performed parallel and in the vicinity of the EELS line scan using the same measuring parameters with respect to the dispersion, step size and length of the scan. Then the energy scale of each measured EEL spectrum of both line scans was calibrated according to the routine described above.

In the next step the background of the spectra containing the N K edge was fitted by an inverse power law function (see Eq. (6.59)) [170] and subtracted from the measured spectra.

$$B_j = A E_j^{-r} \qquad \qquad (6.59)$$

where j denotes the channel number of the CCD, A and r are constants, B_j is the background intensity and E_j is the energy loss in the corresponding channel number.

The measured EEL spectra are often affected by plural inelastic scattering effects. This means that a transmitted electron will be inelastically scattered more than once. Thus, an incident fast electron that has undergone inner-shell scattering can cause also outer-shell excitation. This mixed inelastic scattering results in an energy loss which is the sum of both processes. It results in a broad peak that is shifted by approximately the plasmon energy to higher energy losses with respect to the ionization edge (core edge) that resulted from inner-shell scattering [171]. The plural inelastic scattering effects can be removed by a deconvolution of the measured spectrum containing the core edge (resulting from inner-shell scattering) with the low loss spectrum. In order to do this for the experiments done within this work, the EEL spectra containing the core edge (in this case the N K edge) were deconvolved with the EEL spectra containing the ZL peak and the low loss region, which were measured along a line scan in the vicinity and parallel to the EELS line scan of the core edge. The deconvolution was done with the Fourier ratio deconvolution routine, see e.g. Ref. [170], implemented within the *DigitalMicrograph^{TM}* software from Gatan [160].

In that way, EEL spectra were obtained which revealed the electron loss near edge structure of the nitrogen absorption edge.

EEL reference spectra of the N K edge for fcc-AlN, w-AlN and fcc-TiN

Due to different features of the near edge structure of the nitrogen edge of the fcc-AlN, w-AlN and fcc-TiN phase, these three phases can be distinguished by EELS. This is illustrated in Fig. 6.7 showing the experimentally obtained spectra of fcc-AlN, w-AlN and fcc-TiN [166] and the calculated spectra of fcc-AlN and w-AlN [167]. These spectra were taken as reference spectra for the analysis of the EEL spectra measured in the Ti-Al-N / Al-Ti-(Ru)-N multilayers and their origin is as follows:

The spectrum of **fcc-AlN** (Fig. 6.7a) was measured in the sample PK131 [172] that was synthesized by a high pressure experiment in a multianvil press and which was provided by Dr. M. Schwarz from the Institute of Inorganic Chemistry of the TU Bergakademie Freiberg. The fcc phase of AlN was verified by SAED and HRTEM / FFT which indicated the zone axis [110]. For comparison the simulated spectrum of fcc-AlN, as taken from Ref. [167], is also displayed in Fig. 6.7a which agrees in principal with the calculated spectra shown by Mizoguchi *et al.* [173] and Le Bossé *et al.* [174]. Mizoguchi *et al.* [173] proposed the onset of the N K edge of fcc-AlN at 2 eV above the onset of the N K edge of w-AlN. The measured spectrum resembles the one measured by Sennour *et al.* [175]. Especially the features labelled by "A" and "D" in Fig. 6.7a were used to identify fcc-AlN in the Ti-Al-N / Al-Ti-(Ru)-N multilayers.

The spectra of **w-AlN** (Fig. 6.7b) was measured in a flake of a w-AlN single crystal that was produced by the Leibniz Institute for Crystal Growth in Berlin. The zone axis was $[212]_w$ as determined by SAED and HRTEM / FFT. For comparison the simulated spectrum of w-AlN, which represents a spectrum that is averaged over all orientations and which was taken from Ref. [167], is also displayed in Fig. 6.7b. The calculated spectrum, see Ref. [167], and the measured spectrum of w-AlN of this work (Fig. 6.7b) agree with the spectra obtained by other authors e.g. Holec *et al.* [176]. The characteristic features of w-AlN are labelled as "A", "B_1", "B_2", "C" and "D" in Fig. 6.7b. The spectrum of w-AlN shown in Fig. 6.7b was taken as reference spectrum. Within the experiments done in this work, the onset of the N K edge in w-AlN is approx. 2 eV above the onset of the N K edge in fcc-TiN.

The spectrum of **fcc-TiN** (Fig. 6.7c) was measured in an fcc-TiN layer that was grown by magnetron sputtering on a $(00\bar{1})$ oriented fcc-MgO single crystal (for details see Section 7.3.1). The zone axis in the fcc-TiN layer was $[0\bar{1}0]_{fcc}$. The measured spectrum of fcc-TiN agrees with the measurement and simulation done by Holec *et al.* [176]. The characteristic features of fcc-TiN are labelled as "A" and "B" in Fig. 6.7c. The energy scale of the EEL spectra was calibrated using the calibration factor f_{EELS} (see Eq. (6.58)) and the position of the Ti $L_{2,3}$ edge, which was assumed to be 455.5 eV according to Ref. [168]. Applying this calibration procedure, the onset of the measured N K edge in fcc-TiN is 395 eV.

The **effect of the crystal orientation of w-AlN** on the near edge structure of the N K edge, which arises from the anisotropy in the electronic structure, is illustrated in Fig. 6.7d. Using the same experimental conditions for the aquistion of the EEL spectra, namely energy of the fast electrons equal to 200 keV, $\alpha_{conv} = 22$ mrad, $\beta_{coll} = 5.5$ mrad, changes in the intensity of the characteristic features "A", "B_1" and "B_2" became apparent in the EEL spectra of w-AlN if the incident electron beam direction is parallel to the $[001]_w$ direction (see Fig. 6.7d) as compared to the incident electron beam direction being parallel to the $[100]_w$ direction (see Fig. 6.7e). In case of the incident electron beam direction being parallel to the $[001]_w$ direction, the features "A", "B_1" and "B_2" are visible, but the intensity of the feature "B_1" increases and gets larger than the feature "B_2". This change in the intensity of the feature "B_1" could be observed experimentally and also in the simulated spectrum in Ref. [167] (see Fig. 6.7d).

Radtke *et al.* [177] observed a similar behaviour in his experiments when the electron beam direction was parallel to the $[001]_w$ direction and a different contribution of the momentum transfer parallel ($q_∥$) and perpendicular ($q_⊥$) to the optical axis (which corresponds to the c axis in his case), as illustrated in Fig. 6.8, was selected by the size of the collection angel β_{coll}.

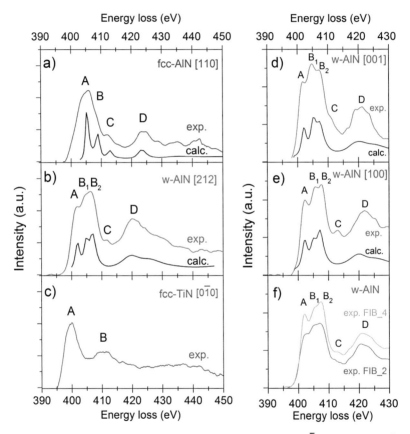

Fig. 6.7: Measured and calculated EEL spectra of fcc-AlN ($Fm\bar{3}m$) (a), w-AlN ($P6_3mc$) (b + d + e + f) and fcc-TiN ($Fm\bar{3}m$) (c) revealing the near edge structure of the nitrogen K edge. The directions given in square brackets that are shown in figures (a) to (e) correspond to the zone axis of the measured phase. The calculated spectra for fcc-AlN ($Fm\bar{3}m$) and w-AlN ($P6_3mc$) were taken from Ref. [167]. The calculated spectra in figure (d) and figure (e) refer to a distinct beam direction, namely $[001]_w$ and $[100]_w$, respectively. The other calculations are averaged over all directions. Significant features in the spectra are labelled by A, B, B_1, B_2, C, and D.

In case of the N K edge in w-AlN, the unoccupied p states in the conduction band of the nitrogen atoms determine the near edge structure of the N K edge [178]. According to Radtke *et al.* [177] the p_z orbitals are excited, if the momentum transfer is parallel to the c axis. If the momentum transfer is perpendicular to the c axis, then (p_x, p_y) orbitals are excited [177]. Regarding the density of states projected on the $p_x + p_y$ states and p_z states, as shown e.g. in Refs. [177, 178], the increased intensity of the feature "B_1" in the measured ELNES in Fig. 6.7d can be attributed

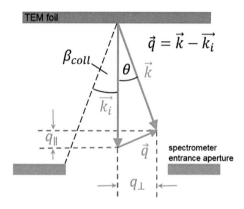

Fig. 6.8: Scattering diagram illustrating the scattering angle θ, the incident electron momentum $\vec{k_i}$ and the momentum transfer $\vec{q}$ with the corresponding parallel $q_{\parallel}$ and perpendicular $q_{\perp}$ components of the momentum transfers.

to the dominance of the $p_x + p_y$ states due to a significant influence of the momentum transfer perpendicular to the c axis as a results of the large size of the used collection angle ($\beta_{coll} = 5.5$ mrad). In the spectra which were acquired with an electron beam direction different than $[001]_w$ (see Fig. 6.7b, Fig. 6.7e and Fig. 6.7f), the higher intensity of the feature "B_2" can be attributed to the significant influence of the p_z states.

Integral intensity of the N K edge

In case of TiN, the relative composition of nitrogen and titanium can be estimated from the analysis of the integral intensity of the N K edge and the Ti $L_{2,3}$ edge using the quantification routine implemented in the *DigitalMicrograph^{TM}* software from Gatan [160].

The determination of the integral intensity of the N K edge is not only valuable for the determination of the concentrations of Ti and N in TiN, it provides also information about the phase composition. Because recently it could be shown that on the basis of the integral intensity of the N K edge, fcc-TiN, fcc-AlN / w-AlN and fcc-Ti$_{1-x}$Al$_x$N can be distinguished [166]. For this analysis an EELS line scan is required that includes measuring points in fcc-TiN, which is used as reference, and in the unknown sample area, which contains either fcc-TiN, fcc-

$Ti_{1-x}Al_xN$ or fcc-AlN / w-AlN. After correction of the measured spectra (energy scale, plural scattering and background, as described above), the intensity of the N K edge and if present the intensity of the Ti $L_{2,3}$ edge is determined using the EELS quantification routine implemented in the *DigitalMicrograph*^{TM} software from Gatan [160]. In case of the presence of both edges, the integration width is ~ 30 eV. Additionally, the relative composition of Ti and N on the fcc-TiN reference site is determined with the EELS quantification routine [160]. If the relative composition of Ti and N measured on the reference yields 50 %, the thickness of the TEM foil along the whole line scan is considered to be suitable for further analysis. If this prerequisite is fulfilled, the integral intensity of the N K edge for fcc-TiN (I_{TiN}^N) is determined and can be used as reference. Then the ratio of the integral intensity of the N K edge of the unknown phase ($I_{unknown}^N$) and the reference (I_{TiN}^N) is calculated. This ratio indicates if the unknown phase is fcc-TiN, fcc-$Ti_{1-x}Al_xN$ ($0.4 \leq x \leq 0.65$) or fcc-AlN / w-AlN because it was found in Ref. [166] that I_{TiN}^N/I_{TiN}^N equals 1, $I_{Ti_{1-x}Al_xN}^N/I_{TiN}^N$ is in the range of 1.4 to 1.5 for $0.4 \leq x \leq 0.65$ and I_{AlN}^N/I_{TiN}^N is in the range of 2.4 to 2.6 (for w-AlN or fcc-AlN).

Determination of the TEM foil thickness

From the spectra containing the ZL peak the thickness t of the TEM foil was determined. The relative thickness of the TEM foil given in relative units of the mean free path (MFP) for inelastic scattering was computed with the *DigitalMicrograph*^{TM} software from Gatan [160]. The computation routine [179] separates the ZL peak from the spectrum and integrates the ZL counts to obtain the intensity I_0. After that the intensity I_t was determined that corresponds to the total area under the whole spectrum up to 2 keV. For this purpose the background of the measured loss part of the spectrum is extrapolated to 2 keV energy loss with an inverse power law function. Finally, the relative thickness t/λ_{MFP} in units of inelastic mean free path was computed according to Eq. (6.60) [171]:

$$t/\lambda_{MFP} = \ln\left(I_t/I_0\right) \tag{6.60}$$

where λ_{MFP} is the total mean free path for all inelastic scattering.

In order to determine the absolute thickness t of the TEM foil, λ_{MFP} was calculated according to Eq. (6.61) [156] [171].

$$\lambda_{MFP} \approx \frac{106 \cdot F \cdot (E_0/E_m)}{\ln(2 \cdot \beta_{coll}^* \cdot E_0/E_m)} \tag{6.61}$$

In this equation λ_{MFP} is given in nm, the incident electron energy E_0 in keV, the average energy loss E_m in eV and the effective collection semi angle β_{coll}^* in mrad. The symbol F stands for the relativistic correction factor which corresponds to 0.618 for E_0=200 keV. The effective collection semi angle β^*_{coll} was determined with the program CONCOR2 from Egerton [171]. The average energy loss E_m of the sample was calculated according to Eq. (6.62):

$$E_m = 7.6 \cdot Z_{eff}^{0.36}$$

(6.62)

where Z_{eff} is the effective atomic number. The parameter Z_{eff} was determined with the *Auto-Zeff* software [180]. In case of the investigated Ti-Al-N / Al-Ti-Ru-N multilayers of series III with the global chemical composition $Ti_{0.45}Al_{0.53}Ru_{0.02}N$ Z_{eff} equals 13.41 whereas Z_{eff} for TiN is 15.45 and Z_{eff} for AlN is 10.07 [180]. In our standard experimental setup for the thickness measurement of Ti-Al-N / Al-Ti-Ru-N multilayers of series III the mean free path λ_{MFP} corresponds to 110 nm as calculated according to Eq. (6.61) where $\beta^*_{coll} \approx 22$ mrad, $F = 0.618$ and E_0=200 keV. The mean free path λ_{MFP} for TiN and AlN correspond to 106 nm and 120 nm, respectively. Thus, the absolute TEM foil thickness was estimated using Eq. (6.63):

$$t = \lambda_{MFP} \cdot \ln \left({I_t}/{I_0} \right)$$

(6.63)

6.4   Thermal treatment

The hardness and microstructure changes of the Ti-Al-N / Al-Ti-(Ru)-N multilayer coatings were studied after thermal treatment at 450°C, 650°, 850°C and 950°C. In order to avoid oxidation, the coatings deposited on cemented carbide substrates were inserted into silica glass tubes. Pieces of Ti sponge were added as getter material for oxygen into the tubes. The gas pressure in the tubes was reduced to $2 \cdot 10^{-4}$ Pa, filled with Ar gas to $5 \cdot 10^4$ Pa and sealed with glass. The coatings within the sealed silica glass tubes were annealed of 60 min in a tube furnace from GERO at the requested temperature. The samples were analysed by *ex situ* GAXRD experiments and nanoindentation after each annealing step and thereafter subjected to the next annealing step.

6.5   Calotest

The thickness of the investigated coatings was determined from a Calotest using the device from *CSM Instruments*. During this test a rotating hardened steel sphere with a radius R of 15 mm was pressed on the coating's surface. An abrasive diamond paste with a grain size of 2 - 10 μm was applied to the rotating sphere which consequently creates a crater into the coating.

Due to the abrasion of the coating and the substrate two concentric circles were obtained. The diameter of the inner circle d_i and outer circle D_o could be determined by optical microscopy. The coating thickness d_L was calculated according to Eq. (6.64) [181].

$$d_L = \sqrt{R^2 - \frac{d_i^2}{4}} - \sqrt{R^2 - \frac{D_o^2}{4}} \qquad \textbf{(6.64)}$$

6.6   Nanoindentation

Instrumented indentation tests were used to determine the hardness and indentation modulus of the coatings according to the Oliver and Pharr method [182, 183]. This kind of test allows to measure the hardness and modulus at a low penetration depth, which should be less than 10 % or even 5 % of the coating thickness [131], so that effects of the deformed substrate on the hardness can be avoided. Due to the low penetration depth the size of the remaining indent is too small to be measured optically. Therefore, load and indentation depth are recorded continuously during the stepwise loading and unloading of the sample (see Fig. 6.9a).

At the maximum load the maximum penetration h_{max} is reached which is the sum of the surface displacement h_s and the contact depth h_c as illustrated in Fig. 6.9b and by Eq. (6.65).

$$h_{max} = h_c + h_s \qquad \textbf{(6.65)}$$

After the load removal the elastic displacements are recovered and the residual imprint has the depth h_0. The indentation hardness H_{IT}, which represents the resistance to permanent deformation [183], is calculated from the maximum load F_{max} and the projected contact area A_p (see Eq. (6.66)).

$$H_{IT} = \frac{F_{max}}{A_p} \qquad \textbf{(6.66)}$$

The contact area A_c is determined by the depth of contact h_c and the indenter geometry. In case of a Berkovich indenter A_p is calculated from the depth of contact h_c according to Eq. (6.67):

$$A_p = 23.96 \cdot h_c^2 + \sum_{i=0}^{7} C_i \cdot h_c^{1/2^i} \qquad \textbf{(6.67)}$$

where C_i are constants describing the deviation from a perfect tip shape due to blunting. The constants C_i where determined regularly by a calibration sequence on fused silica with known elastic modulus.

In order to obtain the contact depth h_c from equation (6.65), the surface displacement h_s has to be determined from Eq. (6.68):

$$h_s = \epsilon \cdot \frac{F_{max}}{S}$$

(6.68)

where ϵ is a geometric constant that is equal to 0.75 for a Berkovich indenter [183] and S is the contact stiffness. The contact stiffness is derived from the upper portion of the unloading curve (see Fig. 6.9a). This part is described by a power law function according to the Oliver and Pharr method [182] and its derivative yields the contact stiffness S (see Eq. (6.69)):

$$S = \frac{dF}{dh}\bigg|_{h=h_{max}}$$

(6.69)

Knowing the contact stiffness S and the projected contact area A_p the so called reduced modulus E_r can be calculated from Eq. (6.70):

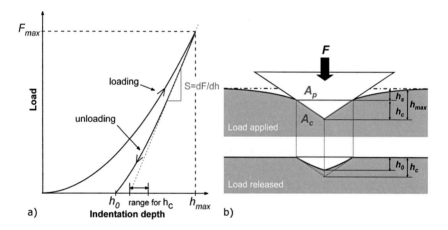

Fig. 6.9: Schematic presentation of the load-displacement curve (a) and the hardness indent with a conical indenter (b) reproduced after [182] and [184], respectively. The upper limit of h_c in (a) corresponds to a conical indenter with the geometric constant $\epsilon = 0.72$ and the lower limit to a flat punch with $\epsilon = 1$. The maximum penetration depth h_{max}, the depth of surface displacement h_s, the contact depth h_c, the depth of the residual imprint h_0, the contact area A_c and projected contact area A_p are illustrated in (b).

$$E_r = \frac{S}{2}\sqrt{\frac{\pi}{A_p}}$$

(6.70)

The reduced modulus corresponds to the combined elastic behaviour of the sample and the indenter material. In order to obtain the indentation modulus of the sample, the elastic modulus of the indenter E_i and the Poisson's ratio of the indenter v_i and the sample v have to be inserted into Eq. (6.71). In case of a diamond indenter E_i equals 1070 GPa and v_i is 0.07.

$$E_{IT} = \frac{1 - \nu^2}{\dfrac{1}{E_r} - \dfrac{1 - \nu_i^2}{E_i}}$$

(6.71)

The indentation experiments were done on a locally polished surface area using a nanohardness tester from CSM Instruments that was equipped with a diamond Berkovich indenter. The maximum load was applied within 30 s and removed within 30 s. In order to keep the penetration depth less than 10 % of the coating thickness, the maximum load lay in the range of 30 to 60 mN. The load displacement curves were analysed using the Oliver and Pharr method [182]. The average coating hardness and indentation modulus were determined from at least 20 indentations for each sample.

7 Results and discussion

7.1 Ti-Al-N coatings deposited by CAE

The influence of the Al content and the substrate bias voltage applied during the deposition of Ti-Al-N coatings by CAE on their microstructure, hardness and thermal stability is shown in this Section. The results were published in References [83, 93].

7.1.1 Deposition of CAE Ti-Al-N monolayer coatings

The Ti-Al-N monolayer coatings were produced in a CAE deposition process. The deposition was done by Ceratizit Luxembourg S.à.r.l. using an industrial CAE facility of the Balzers RCS type [185] and the standard deposition program "Futura Nano". The deposition was done from four circular powder metallurgical (PM) Ti-Al targets from Plansee CM [186]. A schematic cross-section of the deposition chamber is shown in Fig. 7.1. The targets were mounted on position 1, 4, 5 and 6 in the deposition chamber.

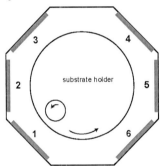

Fig. 7.1: Schematic drawing of the Balzers RCS deposition unit.

The diameter of the targets was 160 mm and the original thickness was 12 mm. The different coatings' compositions were achieved via four different target compositions: (I) 60 at.% Ti and 40 at.% Al, (II) 50 at.% Ti and 50 at.% Al, (III) 40 at.% Ti and 60 at.% Al and (IV) 33 at.% Ti and 67 at.% Al. In order to modify the ion impact during the deposition process, the four depositions were done at four different bias voltages (U_B): -20 V, -40 V, -80 V and -120 V. In this way a matrix of 4 × 4 samples was produced. Polished cemented carbide inserts containing 12 wt.% Co and mixed carbides (hex-WC, fcc-TiC) with SNUN 120412 geometry according to DIN ISO 1832 [187] were used as substrates. The cemented carbide inserts were magnetically mounted on the substrate holder. In order to improve the adhesion of the coatings, prior to the deposition of the $Ti_{1-x}Al_xN$ layer a TiN base layer with a thickness of approx. 0.2 μm was deposited using two PM Ti targets which were mounted on position 2 and 3. A sketch of the coating architecture is shown in Fig. 7.18a. The deposition was done using two-fold rotation at 450 °C in nitrogen atmosphere at a working pressure of 3.2 Pa. The deposition time for the $Ti_{1-x}Al_xN$ layer was 114 min.

7.1.2   Chemical and phase composition of as-deposited $Ti_{1-x}Al_xN$ coatings

The analysis of the chemical composition of the as-deposited Ti-Al-N coatings done by EPMA / WDX and GDOES (see Sections 6.1.1 and 6.1.2), revealed that the actual [Al] / ([Ti]+[Al]) ratio in the coatings is lower than in the mixed Ti-Al targets that were used for the deposition. This effect is attributed to a higher re-sputtering yield of already deposited Al on the substrate as discussed in Section 3.2.2.1 and is in accordance with References [59-62]. However, no preferential incorporation of Ti in the coatings was observed with increasing bias voltage. The actual composition of the four coating series can be summarized as follows: $Ti_{0.62\pm0.01}Al_{0.38\pm0.01}N_{1\pm0.05}$, $Ti_{0.53\pm0.01}Al_{0.47\pm0.01}N_{1\pm0.05}$, $Ti_{0.44\pm0.01}Al_{0.56\pm0.01}N_{1\pm0.05}$ and $Ti_{0.38\pm0.01}Al_{0.62\pm0.01}N_{1\pm0.05}$.

The phase composition of the coatings could be concluded from the intensity of the diffraction lines corresponding to the expected crystalline phases fcc-(Ti,Al)N and w-AlN. Each diffraction line was fitted by a symmetrical Pearson VII function as shown in Fig. 7.2. Additionally, the stress-free lattice parameter of the fcc-(Ti,Al)N phase was determined as shown in Fig. 7.3a. The comparison of the stress-free lattice parameters of the fcc-(Ti,Al)N phase with the assumed dependence of the lattice parameter of fcc-$Ti_{1-x}Al_xN$ on the Al content according to Eq. (7.1) [92] yielded further information on the phase composition.

$$a(Ti_{1-x}Al_xN) = [0.42418(2) - 0.01432(2) \cdot x] \text{ nm} \qquad (7.1)$$

Eq. (7.1) represents a Vegard-like dependence for the fcc-$Ti_{1-x}Al_xN$ phase which indicates whether the whole Al content of a Ti-Al-N coating is incorporated into the crystalline fcc-$Ti_{1-x}Al_xN$ phase present in that coating. For this purpose the stress-free lattice parameter $a(Ti_{1-x}Al_xN)$ for the average Al content present in the whole coating is calculated using Eq. (7.1). If the calculated stress-free lattice parameter (a_0) agrees with the measured a_0 of the fcc-(Ti,Al)N phase, then the whole Al content present in the coating is incorporated into the fcc-(Ti,Al)N phase. If the calculated stress-free lattice parameter (a_0) is higher than the measured one, then the Al content in the majority of the fcc-(Ti,Al)N phase is below the average Al content of the coating. The remaining Al that is not incorporated into the majority of the fcc-

(Ti,Al)N phase could be present in another phase. This phase could be the wurtzite phase, the fcc-AlN phase or an Al-rich fcc-(Al,Ti)N phase or even an amorphous phase.

Despite the Al content, also the nitrogen content affects the lattice parameter of the fcc-$Ti_{1-x}Al_xN_{1\pm\delta}$ phase [188, 189]. The experimental work of Nagakura *et al.* [188] showed that a nitrogen deficiency of $\delta = 0.05$ in $TiN_{1-\delta}$ would lead to a decrease of the fcc-TiN lattice parameter by 0.0002 nm. According to *ab initio* calculations by Baben *et al.* [189] a nitrogen overstoichiometry of $\delta = 0.033$ in fcc-$Ti_{0.5}Al_{x0.5}N_{1+\delta}$ as caused by metal vacancies would lead to a decrease of the lattice parameter by 0.0004 nm. However, since the uncertainty in the nitrogen stoichiometry is approx. ± 0.05 in the coatings investigated within this work, the effect of the departure from the Vegard-like dependence is low and is not considered in the following discussions.

The phase analysis based on the presence of diffraction lines in the GAXRD pattern and on the analysis of the stress-free lattice parameter of the fcc-(Ti,Al)N phase revealed that the phase composition of the coatings with the same chemical composition is different for the coatings deposited at low bias voltage (-20 V and -40 V) and high bias voltage (-80 V and -120 V). The phase compositions of all coatings are summarized in Fig. 7.3b. The $Ti_{0.62}Al_{0.38}N$ and $Ti_{0.53}Al_{0.47}N$ coatings deposited at a bias voltage of $U_B = $ -20 V (Fig. 7.2a) and $U_B = $ -40 V (Fig. 7.2b) contained fcc-(Ti,Al)N as single crystalline phase. This is supported by the analysis of the stress-free lattice parameter of the fcc-(Ti,Al)N phase which is close to the stress-free lattice parameter expected for the chemical composition of the $Ti_{0.62}Al_{0.38}N$ and $Ti_{0.53}Al_{0.47}N$ coatings according to Eq. (7.1). When the Al content was increased to $x = 0.56$ at low bias voltages, the formation of traces of w-AlN was already visible in the GAXRD patterns (see Fig. 7.2a and Fig. 7.2b). Metastable fcc-(Ti,Al)N was the major phase. From the analysis of the GAXRD pattern obtained with synchrotron radiation of the $Ti_{0.44}Al_{0.56}N$ coating deposited at $U_B = $ -40 V, the estimated phase fraction of the wurtzite phase was approx. (14 ± 5) vol.% or (12 ± 5) mol %. A further increase of the Al content to $x = 0.62$ and the application of a low substrate bias voltage yielded a decrease of the fcc-(Ti,Al)N phase fraction whereas w-AlN became the major crystalline phase. The diffraction lines of the wurtzite phase in the coatings were shifted from the expected 2θ-positions of w-AlN with the lattice parameters $a = 0.311$ nm and $c = 0.498$ nm [190] which indicated the expansion of the wurtzite elementary cell due to the incorporation of Ti into the wurtzite phase [191].

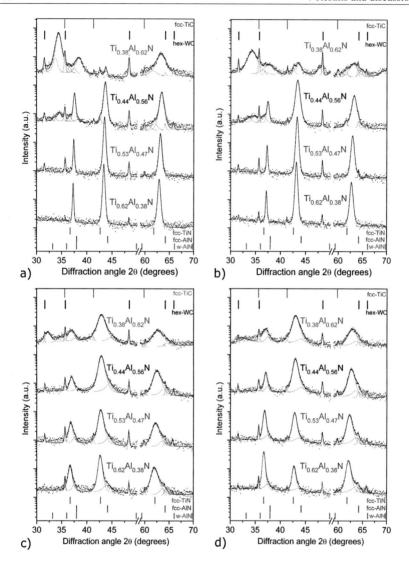

Fig. 7.2: Parts of the GAXRD patterns measured for the Ti$_{0.62}$Al$_{0.38}$N, Ti$_{0.53}$Al$_{0.47}$N, Ti$_{0.44}$Al$_{0.56}$N and Ti$_{0.38}$Al$_{0.62}$N coatings that were deposited on cemented carbide substrates at bias voltages of -20 V (a), -40 V (b), -80 V (c) and -120 V (d). The points represent the measured data; grey lines represent the individual diffraction lines as fitted by the Pearson VII function; black lines are the intensities from the whole fitted profile. The positions of fcc-TiN, fcc-AlN and w-AlN diffraction lines are labelled at the bottom of the figure and the positions of hex-WC and fcc-TiC diffraction lines coming from the substrate are shown at the top of the figure.

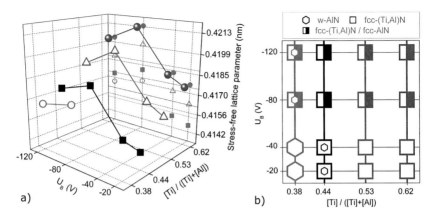

Fig. 7.3: Influence of the bias voltage and the [Ti] / ([Ti]+[Al]) ratio in the coatings on the stress-free lattice parameter of the fcc-(Ti,Al)N phase (a) and the phase composition of the $Ti_{1-x}Al_xN$ coatings (b).

In contrast to the Ti-Al-N coatings deposited at low bias voltages, fcc-(Ti,Al)N was the major phase in all investigated coatings deposited at high bias voltages (-80 V and -120 V). The application of a high bias voltage seems to suppress the formation of w-AlN, since the diffraction lines of w-AlN appear for the first time in the GAXRD patterns of the $Ti_{1-x}Al_xN$ coatings deposited at high U_B at the highest Al content of x = 0.62 (see Fig. 7.2c and Fig. 7.2d). Only traces of w-AlN were found in these coatings whereas w-AlN was the dominating phase in the $Ti_{1-x}Al_xN$ coatings deposited at low bias voltage (see Fig. 7.3b). The effect of the reduction of the volume fraction of the w-AlN phase at high bias voltages was previously observed by Pfeiler *et al.* in $Ti_{1-x}Al_xN$ based coatings that contained vanadium [192] and tantalum [193]. Recently, this effect could be confirmed in another Al-rich $Ti_{1-x}Al_xN$ coating series with $0.61 \leq x \leq 0.64$ [191].

As mentioned above, in the $Ti_{1-x}Al_xN$ coatings with $0.38 \leq x \leq 0.62$ deposited at high bias voltages (-80 V and -120 V) the major phase was fcc-(Ti,Al)N. However, the diffraction lines of fcc-(Ti,Al)N were characterized by a slight asymmetry with higher intensities of the right-hand tails and could only be fitted by two symmetric Pearson VII functions (see Fig. 7.2c and Fig. 7.2d). The asymmetry was especially pronounced for the 200 diffraction line. The same asymmetry of the fcc-(Ti,Al)N diffraction lines was observed in the GAXRD patterns of $Ti_{0.54}Al_{0.46}N$ coatings deposited onto (111) oriented Si substrates at a bias voltage of $U_B = -80$ V (see Fig. 7.4). This means that the asymmetry originates from the coating itself since an overlap of the 200 diffraction line with a diffraction line originating from the substrate that could cause this asymmetry can be excluded in case of Si substrates. Therefore, this asymmetry was interpreted as an overlap of fcc-(Ti,Al)N with weak broad diffraction lines from Al-rich fcc-(Al,Ti)N, which is labeled as fcc-AlN for simplicity. The overlap of fcc-(Ti,Al)N and fcc-AlN explains also the different extent of the asymmetry as function of the diffraction indices, since the diffraction lines with odd diffraction indices e.g. 111 show a less asymmetric peak shape than the diffraction lines with even diffraction indices hkl like 200 and 220 (cf. Fig. 7.4). This effect, which is explained in Ref. [48], is attributed to the magnitude of the structure factor F_{hkl} for fcc-Ti$_{1-x}$Al$_x$N that is a function of hkl and the Al concentration x. A brief derivation of the intensity ratios of fcc-AlN and fcc-Ti$_{1-x}$Al$_x$N is given in the following.

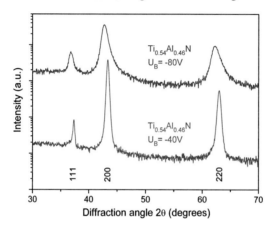

Fig. 7.4: Low angle parts of the GAXRD patterns of the Ti$_{0.54}$Al$_{0.46}$N coatings deposited onto (111) oriented Si substrate at U$_B$= -40 V and U$_B$= -80 V.

The integral intensities ($I(hkl)$) of diffracting fcc-Ti$_{1-x}$Al$_x$N domains are proportional to the squared magnitude of the structure factor. Assuming that the nitrogen atoms are located at the lattice positions (½, 0, 0), (0, ½, 0), (0, 0, ½) and (½, ½, ½) and the Al and Ti atoms occupy the lattice positions (0, 0, 0), (½, ½, 0), (½, 0, ½) and (0, ½, ½) in the fcc-Ti$_{1-x}$Al$_x$N phase, the squared structure factor ($|F_{hkl}|^2$) for diffraction lines with even and odd hkl indices corresponds to Eq. (7.2) and Eq. (7.3), respectively.

$$I(hkl)_{even} \sim |F_{hkl}|^2 = 16 \cdot |x \cdot f_{Al} + (1-x) \cdot f_{Ti} + f_N|^2 \qquad \text{(7.2)}$$

$$I(hkl)_{odd} \sim |F_{hkl}|^2 = 16 \cdot |x \cdot f_{Al} + (1-x) \cdot f_{Ti} - f_N|^2 \qquad \text{(7.3)}$$

where f_{Al}, f_{Ti} and f_N are the atomic scattering factors of Al, Ti and N, respectively. Equations (7.2) and (7.3) were obtained from the general definition of the structure factor (see Eq. (7.4)) which corresponds to Eq. (7.5) for metal nitride with NaCl crystal structure.

$$F(hkl) = \sum_{j=1}^{N} f_j \cdot \exp[2\pi i(hx_j + ky_j + lz_j)] \tag{7.4}$$

$$F(hkl) = f_{Me} \cdot \{\exp(0) + \exp[\pi i(h+k)] + \exp[\pi i(h+l)] + \exp[\pi i(k+l)]\}$$
$$+ f_N \cdot \{\exp[\pi i(h+k+l)] + \exp(\pi ih) + \exp(\pi ik) + \exp(\pi il)\} \tag{7.5}$$

where f_{Me} is the atomic scattering factor of the metal and (x_j, y_j, z_j) are the lattice positions which equal (½, 0, 0), (0, ½, 0), (0, 0, ½), (½, ½, ½) for N and (0, 0, 0), (½, ½, 0), (½, 0, ½), (0, ½, ½) for the metal sites. The atomic scattering factors f_{Me} and f_N were calculated from the parameters of the analytical scattering-factor functions (a_i, b_i, c) given by Waasmaier *et al.* [194] according to Eq. (7.6):

$$f_{Me,N} = \sum_{i=1}^{4} a_i \exp\left(-b_i \cdot \left(\frac{\sin\theta}{\lambda}\right)^2\right) + c \tag{7.6}$$

The ascending order of the atomic scattering factors is $f_N < f_{Al} < f_{Ti}$. Thus, an increasing Al content in fcc-$Ti_{1-x}Al_xN$ leads to a decrease of the diffracted intensity which is stronger for diffraction lines with odd indices than for diffraction lines with even indices. The ratios of the squared magnitudes of the structure factors of fcc-AlN and fcc-$Ti_{0.5}Al_{0.5}N$ correspond to the ratio $I_{hkl}(AlN)/I_{hkl}(Ti_{0.5}Al_{0.5}N)$ and yielded 0.330, 0.640 and 0.657 for the diffraction lines 111, 200 and 220, respectively. This ratio is nearly twice for the diffraction line 200 and 220 as compared to 111. Thus, the asymmetry is more visible for the 200 and 220 diffraction lines than for the 111 diffraction line.

Additionally to the observed asymmetry, the stress-free lattice parameter of the fcc-(Ti,Al)N phase of the coatings deposited at high U_B was larger than the expected lattice parameter according to the Vegard-like dependence (Eq. (7.1)). The co-existence of two fcc phases having a stress-free lattice lattice parameter below and above the expected Vegard-like dependence can be considered as a sign for the presence of fluctuations in the Ti and Al concentrations in fcc-(Ti,Al)N which were observed by atom probe experiments [195]. Hence, it can be concluded that the $Ti_{0.62}Al_{0.38}N$, $Ti_{0.53}Al_{0.47}N$ and $Ti_{0.44}Al_{0.56}N$ coatings deposited at high bias voltages consist of fcc-(Ti,Al)N which has the Al content below the nominal Al ratio of the coating and an Al-rich fcc-(Al,Ti)N which contains nearly no Ti and is therefore labelled as fcc-AlN in Fig. 7.3b. The presence of metastable fcc-AlN in the as-deposited coatings could be facilitated by local lattice strains at the fcc-(Ti,Al)N / fcc-AlN interfaces which was proven experimentally in Ref. [82].

In contrast to the $Ti_{1-x}Al_xN$ coatings deposited at high bias voltages ($U_B = -80$ V and -120 V), the single phase nature of the $Ti_{1-x}Al_xN$ coatings whith $x < 0.44$ which were deposited at low bias voltages ($U_B = -20$ V and -40 V) indicates that the Ti and Al atoms in (Ti,Al)N are distributed more homogeneously or rather fluctuate more shallowly.

7.1.3   Macroscopic lattice strain and residual stress of as-deposited $Ti_{1-x}Al_xN$ coatings

The macroscopic lattice strain (ε), which was determined from the dependence of the measured lattice parameter on $\sin^2\psi$ and calculated according to Eq. (6.23), is shown for the fcc-(Ti,Al)N phase in Fig. 7.5. The Poisson's ratio used for the calculation was assumed to be 0.3 [196]. Supposing a Young's modulus of 500 GPa [82] the macroscopic residual stress of the fcc-(Ti,Al)N phase as determined from Eq. (6.24) is given additionally in Fig. 7.5. Since the dominating phase was w-AlN in the $Ti_{0.38}Al_{0.62}N$ coatings deposited at $U_B = -20$ V and $U_B = -40$ V the residual stress of the fcc-(Ti,Al)N phase was not determined. The analysis of the macroscopic lattice strains and the residual stress has shown that the bias voltage strongly influences the stress state of the coatings.

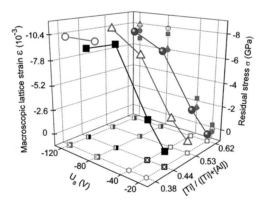

Fig. 7.5: Influence of the bias voltage and the [Ti] / ([Ti]+[Al]) ratio on the macroscopic lattice strain and macroscopic residual stress in the fcc-(Ti,Al)N phase. The lines are guide for the eyes. The symbols in the horizontal plane represent the coatings' phase composition, their meaning corresponds to the legend of Fig. 7.3b. The grey symbols are the projection of the determined values on the vertical plane of the 3-dimensional plot.

The $Ti_{1-x}Al_xN$ coatings with an Al concentration of $x = 0.38$ and $x = 0.47$ that were deposited at the lowest bias voltage $U_B = -20$ V were characterized by low tensile stress (< 0.5 GPa) and by almost zero residual stress for $x = 0.56$. Compressive residual stress was observed in the fcc-

(Ti,Al)N phase of the coatings when the negative bias voltage was increased to $U_B = -40$ V, -80 V and -120 V. The residual compressive stress rose by a factor of 4 in the $Ti_{0.62}Al_{0.38}N$, $Ti_{0.53}Al_{0.47}N$ and $Ti_{0.44}Al_{0.56}N$ coatings when the negative bias voltage was increased from -40 V to -80 V. A further increase of the negative bias voltage to -120 V in the $Ti_{0.62}Al_{0.38}N$ and $Ti_{0.53}Al_{0.47}N$ coatings resulted in slightly higher compressive stress as compared to the coatings deposited at $U_B = -80$ V, whereas in the $Ti_{0.44}Al_{0.56}N$ and $Ti_{0.38}Al_{0.62}N$ coatings the stress saturated at -120 V. The increase of the residual stress with increasing negative bias voltage to $U_B = -80$ V is attributed on the one hand to a higher density of lattice defects e.g. displaced atoms due to direct and recoil implantation of film atoms [66, 197, 198] caused by the increasing kinetic energy of the bombarding ions due to the rising potential difference (ΔV_{sheath}) between the substrate and the plasma potential with increasing U_B, as described by Eq. (2.6). At high bias voltages (e.g. $U_B = -120$ V), the impinging ions deliver some energy to the film surface leading to vibration and generation of phonons in the vicinity of the collision cascade. This facilitates short range movements of atoms and relaxation of displaced atoms [198]. In addition, local heating of the deposited film takes place at a high ion flux leading to an increased annealing rate of created defects [198]. This effect is reflected by minor increase or rather saturation of the residual stress at $U_B = -120$ V.

On the other hand the residual stress in fcc-(Ti,Al)N correlates with the phase composition of the coatings, because in all coatings where high compressive residual stress was observed, fcc-AlN was found as a secondary phase to fcc-(Ti,Al)N (see Fig. 7.3b). This implies that lattice strains are introduced due to a misfit between both phases. Furthermore, under compression the solubility limit of Al in fcc-$Ti_{1-x}Al_xN$ increases as shown by *ab initio* calculations of Holec *et al.* [199].

7.1.4   Crystallite size of as-deposited $Ti_{1-x}Al_xN$ coatings

Large broadening of the diffraction lines of the fcc-(Ti,Al)N phase (see Fig. 7.2) indicated that the $Ti_{1-x}Al_xN$ coatings consist of nanocrystallites. The analysis of the XRD line broadening revealed that the nanocrystallites are partially coherent [200]. The presence of partially coherent nanocrystallites having a small mutual misorientation was observed in various ternary and quaternary TM-Al-(Si)-N (TM = Ti, Cr, Zr) coatings [57, 92, 109, 201]. Partially coherent nanocrystallites result from local epitaxy [82] or the fragmentation of larger crystallites [109]. Such a large crystallite that is fragmented into partially coherent nanocrystallites with a small mutual misorientation is shown in the TEM dark field image in Fig. 7.6a that was taken from the $Ti_{0.53}Al_{0.47}N$ coating deposited at $U_B = -120$ V. The sizes of the fcc-(Ti,Al)N nanocrystallites as determined from the analysis of the line broadening according to the procedure described in Ref. [200] are given in Fig. 7.6b. Additionally, the coatings' phase composition is shown at the bottom of Fig. 7.6b which indicates the correlation of the nanocrystallites with the phase composition of the coatings. The largest fcc-(Ti,Al)N nanocrystallites having a size of $\sim$ 9-11 nm were observed in the single-phase $Ti_{0.62}Al_{0.38}N$ and $Ti_{0.53}Al_{0.47}N$ coatings that were deposited at low bias voltages of $U_B = -20$ V and $U_B = -40$ V. The formation of w-AlN as second phase in the $Ti_{0.47}Al_{0.53}N$ coatings deposited at $U_B = -20$ V and $U_B = -40$ V reduced the size of the fcc-(Ti,Al)N nanocrystallites to $\sim$ 6 nm. When the Al content was increased to $x = 0.62$ in the $Ti_{1-x}Al_xN$ coatings deposited at low bias voltage, w-AlN was the dominating phase and the determination of the size of the fcc-(Ti,Al)N nanocrystallites was not possible. The dual-phase $Ti_{1-x}Al_xN$ coatings with $0.38 \leq x \leq 0.56$ and containing fcc-(Ti,Al)N and fcc-AlN, which were deposited at high bias voltages of $U_B = -80$ V and $U_B = -120$ V, were characterized by substantially smaller crystallites than single-phase coatings. Their size was in the range of $\sim$ 3.5 - 5 nm. The additional formation of w-AlN beside fcc-(Ti,Al)N and fcc-AlN in $Ti_{0.38}Al_{0.62}N$ coatings deposited at high bias voltages yielded a size of the fcc-(Ti,Al)N nanocrystallites of $\sim$ 3 nm.

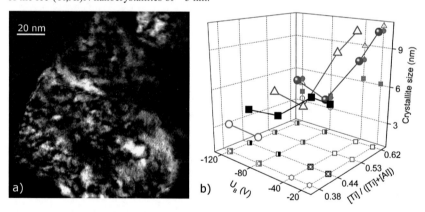

Fig. 7.6: a) TEM dark field image of the $Ti_{0.53}Al_{0.47}N$ coating deposited at $U_B = -120$ V showing a crystallite composed of partially coherent nanocrystallites with small mutual misorientation. b) Influence of the bias voltage and the [Ti] / ([Ti] + [Al]) content on the average crystallite size of the fcc-(Ti,Al)N phase. The lines are guide for the eyes. The symbols in the horizontal plane represent the coatings' phase composition, their meaning corresponds to the legend of Fig. 7.3b. The grey symbols are the projection of the determined values on the vertical plane of the 3-dimensional plot.

7.1.5 Interfaces between fcc-(Ti,Al)N and w-AlN

Interfaces play a key role for the hardness of nanocrystalline materials. According to the reverse Hall-Petch relation, the hardness will decrease if the crystallite size decreases below a critical size ($\sim$ 10 - 20 nm). This effect is attributed to grain boundary sliding. Grain boundary sliding can be suppressed by the formation of strong crystallite boundaries [131]. In nanocomposite materials the strengthening of crystallite boundaries can be achieved by the formation of composites that contain a nanocrystalline and an amorphous phase or by composites that contain two nanocrystalline phase with coherent boundaries [131]. Thus, the formation of (partially) coherent interfaces between fcc-(Ti,Al)N and w-AlN in Ti-Al-N coatings is important.

The orientation relationship between fcc-(Ti,Al)N and w-AlN at their interfaces was studied by HRTEM exemplarily in the $Ti_{0.44}Al_{0.56}N$ coating, that was deposited at $U_B = -40$ V. This coating contained fcc-(Ti,Al)N as major phase (see Fig. 7.3b) with approx. (12 ± 5) mol % of wurtzite phase (see Section 7.1.2). HRTEM images of this sample were analyzed by FFT in order to identify regions with wurtzite and fcc phases. The HRTEM micrograph in Fig. 7.7a shows an area where both phases are in contact. The analysis of the FFTs (Fig. 7.7b,c) obtained from the image sections that are bordered by coloured squares revealed the fcc phase in the middle of the picture being in contact with wurtzitic phase at the lower right corner [202, 203] . The directions of the incident electron beam $\vec{B}_{ebeam}$ was $[001]_{fcc}$ in the fcc phase and $[001]_w$ in the wurtzitic phase. Since the vector pointing from the origin of the diffraction pattern to the diffraction spot $(2\bar{2}0)_{fcc}$ (see yellow arrow in Fig. 7.7b) is parallel to the vector pointing from the origin of the diffraction pattern to the diffraction spot $(0\bar{1}0)_w$ (see orange arrow in Fig. 7.7c), it could be concluded that the lattice planes $(1\bar{1}0)_{fcc}$ and $(0\bar{1}0)_w$ are parallel to each other in real space. Additionally, the vector pointing from the origin of the diffraction pattern to the diffraction spot $(\bar{2}\bar{2}0)_{fcc}$ (see yellow arrow in Fig. 7.7b) is parallel to the vector pointing from the origin of the diffraction pattern to the diffraction spot $(\bar{2}10)_w$ (see orange arrow in Fig. 7.7c). This means the lattice planes $(\bar{1}\bar{1}0)_{fcc}$ and $(\bar{2}10)_w$ are parallel to each other in real space. This mutual orientation relationship (OR) between both phases as found in the HRTEM image in Fig. 7.7a can be described as: $(1\bar{1}0)_{fcc} \parallel (0\bar{1}0)_w$ (or $(\bar{1}\bar{1}0)_{fcc} \parallel (\bar{2}10)_w$) and $[001]_{fcc} \parallel [001]_w$. It is illustrated schematically in Fig. 7.8. The projection shown in Fig. 7.8 is parallel to the plane of the TEM foil. Such an interface between both phases introduces lattice strains due to the lattice misfit between both phases. The lattice misfit depends on the Al content in the fcc-$Ti_{1-x}Al_xN$ phase because it determines the lattice parameter parameter of the fcc phase (see Eq. (7.1)).

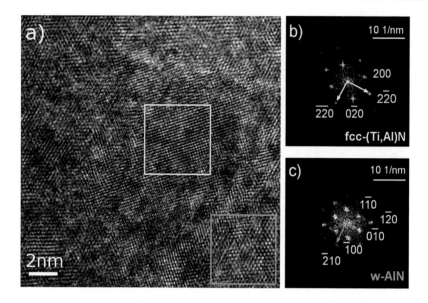

Fig. 7.7: HRTEM micrograph of the $Ti_{0.44}Al_{0.56}N$ coating deposited at $U_B = -40$ V showing the interface of fcc-(Ti,Al)N and w-AlN (a). The FFTs in figures (b) and (c) correspond to the yellow and orange marked square, respectively. The direction of the primary beam is $[001]_{fcc}$ for fcc-(Ti,Al)N and $[001]_w$ for w-AlN. The arrows in figures (b) and (c) represent the diffraction vectors of the parallel lattice planes $(2\bar{2}0)_{fcc}$ and $(0\bar{1}0)_w$ as well as $(\bar{2}\bar{2}0)_{fcc}$ and $(\bar{2}10)_w$.

Considering the interface between wurtzite and fcc-phase that is shown in Fig. 7.8, the distances between the titanium atoms along the direction $[\bar{1}\bar{1}0]_{fcc}$ and the distances of the aluminium atoms along $[\bar{1}00]_w$ are similar. The lattice misfit (f) between the metal atoms along $[\bar{1}\bar{1}0]_{fcc}$ and along $[\bar{1}00]_w$ can be calculated according to Eq. (7.7):

$$f_1 = 2 \cdot \frac{a_{fcc}/\sqrt{2} - a_w}{a_{fcc}/\sqrt{2} + a_w} \tag{7.7}$$

The lattice misfit is negative and ranges between 0.036 and 0.058 for w-AlN with the lattice parameters a = 0.311 nm and c = 0.498 nm [190] and the fcc phases with a lattice parameter ranging from a = 0.424 nm (for fcc-TiN) to a = 0.415 nm (experimentally determined for fcc-(Ti,Al)N in this coating).

As mentioned above the lattice planes $(1\bar{1}0)_{fcc}$ and $(0\bar{1}0)_w$ are also parallel. The lattice misfit calculated from the interplanar spacing between these lattice planes is calculated from Eq. (7.8).

$$f_2 = 2 \cdot \frac{a_{fcc}/\sqrt{2} - a_w \cdot \sqrt{3}/2}{a_{fcc}/\sqrt{2} + a_w \cdot \sqrt{3}/2} \tag{7.8}$$

The misfit is positive and increases from 0.086 to 0.107 when the lattice parameter of the fcc phase increases from a = 0.415 nm to a = 0.424 nm and the lattice parameters of w-AlN are constant as mentioned above.

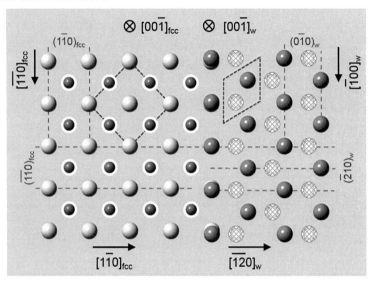

Fig. 7.8: Schematic representation of an fcc-TiN (Fm$\overline{3}$m) and w-AlN (P6$_3$mc) interface with the mutual orientation $(1\overline{1}0)_{fcc} \parallel (0\overline{1}0)_w$ and $[00\overline{1}]_{fcc} \parallel [00\overline{1}]_w$. The crystal structures are projected along the directions $[00\overline{1}]_{fcc}$ and $[00\overline{1}]_w$ which are facing downwards. The elementary cells are marked by the broken blue lines. The interatomic distances within the lattice planes $(1\overline{1}0)_{fcc}$, $(\overline{1}10)_{fcc}$, $(0\overline{1}0)_w$ and $(\overline{2}10)_w$ are indicated by grey dashed lines. The titanium atoms are plotted by yellow spheres, the aluminium by orange spheres and the nitrogen by green spheres. Spheres that are hollow hatched represent atoms which are lying below the plane of the paper.

Fig. 7.9 shows the projection of the mutual orientation between fcc-TiN and w-AlN as seen perpendicular to the TEM foil plane with the $(\overline{1}10)_{fcc}$ and $(\overline{2}10)_w$ planes lying parallel to the normal of the TEM foil and the directions $[1\overline{1}0]_{fcc}$ and $[\overline{1}\overline{2}0]_w$ being orthogonal to the normal of the TEM foil. For a clearer presentation the atoms that are lying below the plane of the paper are shown as hollow hatched spheres. In order to illustrate the degree of the match between both phases, a straight interface is shown at which a row of atoms from both phases overlap in Fig. 7.9. The lattice misfit between the titanium atoms along $[00\overline{1}]_{fcc}$ and the aluminium

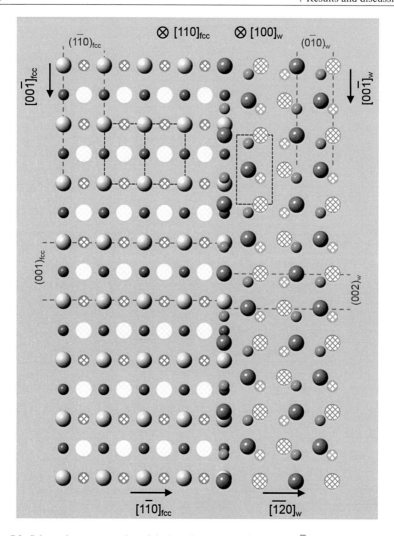

Fig. 7.9: Schematic representation of the interface between fcc-TiN (Fm$\overline{3}$m) and w-AlN (P6₃mc) from Fig. 7.8 and plotted for the projection $(\overline{1}\overline{1}0)_{fcc} \parallel (\overline{2}10)_w$ and $[00\overline{1}]_{fcc} \parallel [00\overline{1}]_w$. The directions $[110]_{fcc}$ and $[100]_w$ are facing downwards. The interatomic distances within the lattice planes $(1\overline{1}0)_{fcc}$, $(001)_{fcc}$, $(0\overline{1}0)_w$ and $(002)_w$ are indicated by grey dashed lines. The elementary cells are marked by the broken blue lines. The titanium atoms are plotted by yellow spheres and the aluminium by orange spheres. The nitrogen atoms originating from fcc-TiN and w-AlN are plotted by dark and bright green spheres, respectively. Atoms that are lying below the plane of the paper are shown as hollow hatched spheres.

atoms along $[00\bar{1}]_w$ or rather the nitrogen atoms along $[00\bar{1}]_{fcc}$ and $[00\bar{1}]_w$ can be calculated according to Eq. (7.9):

$$f_3 = 2 \cdot \frac{a_{fcc} - c_w}{a_{fcc} + c_w} \qquad\qquad (7.9)$$

The misfit is negative and huge. Its magnitude increases from 0.161 to 0.182 when the lattice parameter of the fcc phase decreases from a = 0.424 nm to a = 0.415 nm assuming the above mentioned lattice parameters of w-AlN.

The calculated lattice misfits f_1, f_2 and f_3 suggest that the observed interface in the HRTEM image of Fig. 7.7a is partially coherent.

7.1.6   Hardness of as-deposited Ti$_{1-x}$Al$_x$N coatings

The microstructure of the Ti$_{1-x}$Al$_x$N coatings in terms of phase composition, residual stress and crystallite size, as presented in Sections 7.1.2 to 7.1.5, determines their hardness. Relatively large nanocrystallites (~ 9 - 11 nm) and low compressive stress (< 2 GPa) in single-phase fcc-(Ti,Al)N coatings deposited at low bias voltages (-20 V, -40 V) yielded a low hardness of ~ 25-30 GPa (see Fig. 7.10). The formation of low amounts of w-AlN as second crystalline phase when the Al content is increased to x = 0.56 at low bias voltages (-20 V, -40 V) led to a hardness increase up to ~ 3 GPa. For instance, at $U_B = $ -20 V the hardness increased to (28.6 ± 0.8) GPa in Ti$_{0.44}$Al$_{0.56}$N starting from (25.8 ± 0.7) GPa in Ti$_{0.62}$Al$_{0.38}$N and from (25.1 ± 0.6) GPa in Ti$_{0.53}$Al$_{0.47}$N coatings. In case of $U_B = $ -40 V, the hardness increased from (27.9 ± 0.8) GPa in Ti$_{0.62}$Al$_{0.38}$N and from (29.7 ± 0.9) GPa in Ti$_{0.53}$Al$_{0.47}$N to (31.4 ± 0.9) GPa in Ti$_{0.44}$Al$_{0.56}$N coatings. The hardness increase can be attributed to the reduction of the crystallite size and to a slight increase of the compressive stress as compared to the single phase fcc-(Ti,Al)N coatings. Furthermore, this hardness increase indicates the positive influence of a low amount of w-AlN having partially coherent interfaces with fcc-(Ti,Al)N (see Section 7.1.5) on the hardness as shown e.g. in Refs. [38, 57, 82]. The phase fraction of the wurtzite phase estimated from GAXRD pattern obtained with synchrotron radiation for the Ti$_{0.44}$Al$_{0.56}$N coating deposited at $U_B = $ -40 V was approx. (14 ± 5) vol.% or (12 ± 5) mol %. When w-AlN became the major phase in Ti$_{0.38}$Al$_{0.62}$N coatings that were deposited at low bias voltages the hardness dropped down to (21.4 ± 1.0) GPa for $U_B = $ -20 V and to (23.4 ± 0.6) GPa for $U_B = $ -40 V. This effect is caused by the predominance of the softer w-AlN phase [56].

The highest hardness was observed in the dual phase fcc-(Ti,Al)N / fcc-AlN Ti$_{1-x}$Al$_x$N coatings with $0.38 \leq x \leq 0.56$ that were deposited at high bias voltages (-80 V, -120 V). In these coatings the hardness increased up to 40 % as compared to the single phase fcc-(Ti,Al)N coatings deposited at low bias voltage. This hardness increase is attributed to high compressive residual stress in the range of 6 to 8 GPa and small nanocrystallites with a size of ~ 3.5 - 5 nm. When the Al content was increased to x = 0.62 in the $Ti_{1-x}Al_xN$ coatings that were deposited at U_B = -80 V and U_B = -120 V the hardness dropped below values observed at $0.38 \leq x \leq 0.56$ for U_B = -80 V and U_B = -120 V, respectively, but were still higher than for single-phase coatings. The hardness decreased by nearly 5 GPa from (34.8 ± 0.6) GPa to (29.9 ± 0.4) GPa at a bias voltage of -80 V and by 3 GPa from (36.5 ± 0.6) GPa to (33.5 ± 0.4) GPa at a bias voltage of -120 V. Both $Ti_{0.38}Al_{0.62}N$ coatings contained wurtzite phase in addition to fcc-(Ti,Al)N and traces of fcc-AlN, whereas the estimated volume fraction of the wurtzite phase in the coating deposited at U_B = -80 was approx. (25 ± 10) vol.% and higher than at U_B = -120 V, see Fig. 7.2c and Fig. 7.2d. In both coatings the fcc-(Ti,Al)N phase was characterized by a similar compressive stress (see Fig. 7.5) and similar size of the nanocrystallites (see Fig. 7.6b).

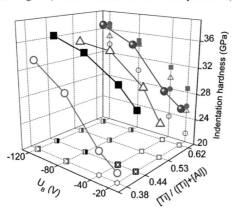

Fig. 7.10: Influence of the bias voltage and the [Ti] / ([Ti]+[Al]) ratio on the indentation hardness of the coatings. The error bars of the individual values are ≤ 1 GPa. The lines are guide for the eyes. The symbols in the horizontal plane represent the coatings' phase composition, their meaning corresponds to the legend of Fig. 7.3b.The grey symbols are the projection of the determined values on the vertical plane of the 3-dimensional plot.

However, the crystallite size was the smallest observed in the whole coating series. A similar behavior was observed for a $Ti_{0.38}Al_{0.62}N$ coating deposited in another CAE process in Ref. [57] that was also characterized by partial coherence as seen from the XRD line broadening, high stress and small crystallite sizes in the range of ~ 3 nm. It is proposed that a possible reason for the hardness reduction are weak interfaces causing increased grain boundary sliding [131]. This is probably promoted by an increased volume fraction of wurtzite phase and its kind of incorporation into the coating. Since the hardness reduction was more pronounced in the $Ti_{0.38}Al_{0.62}N$ deposited at $U_B = -80$ V, which contained a higher volume fraction of wurtzite phase as compared to $Ti_{0.38}Al_{0.62}N$ deposited at $U_B = -120$ V, less partial coherent interfaces between wurtzite and fcc phase could be assumed which diminishes the positive effect of w-AlN to the hardness.

7.1.7   Thermal stability of $Ti_{1-x}Al_xN$ coatings

It was shown in the above Sections (7.1.2 to 7.1.6) that the bias voltage used for the deposition of CAE $Ti_{1-x}Al_xN$ coatings determines their microstructure in the as-deposited state. In order to investigate their thermal stability, *in situ* HT-GAXRD experiments using synchrotron radiation (see Section 6.2.2) were applied. The $Ti_{1-x}Al_xN$ coatings with $0.38 \leq x \leq 0.56$ deposited at $U_B = -40$ V, $U_B = -80$ V and $U_B = -120$ V were annealed and characterized by *in situ* HT-GAXRD at a pressure of $\sim 10^{-3}$ Pa according to the temperature profile shown in Fig. 6.2a. The analysis of the HT-GAXRD patterns revealed a different development of the phase composition as well as the stress-free lattice parameter, macroscopic lattice strain and microstrain of the fcc-(Ti,Al)N phase for the $Ti_{1-x}Al_xN$ coatings with the same [Al] / ([Ti]+[Al]) ratio but deposited at different bias voltages.

The development of the microstructure parameters (phase composition, a_0, ε and e) versus the different annealing steps is presented in the following exemplarily for the $Ti_{0.62}Al_{0.38}N$ coating deposited at $U_B = -40$ V, $U_B = -80$ V and $U_B = -120$ V.

The temperature-composition phase diagram for fcc-$Ti_{1-x}Al_xN$ was described by Mayrhofer *et al.* [204] and Alling *et al.* [205] using *ab initio* calculations. The calculated diagrams show the locus of the spinodal and binodal (see Fig. 7.11). Although the calculated diagrams of both authors are not identical, in both cases the composition $Ti_{0.62}Al_{0.38}N$ is located under the spinodal. Thus the fcc- $Ti_{0.62}Al_{0.38}N$ decomposes through spinodal decomposition into phases at the spinodal. In the case of spinodal decomposition the second derivative of the free energy of mixing (ΔG) with respect to the composition x is negative ($\partial^2 \Delta G/\partial x^2 < 0$). Phases with a composition lying outside the spinodal region decompose via nucleation and growth. In this case the second derivative of the free energy of mixing (ΔG) with respect to the composition x is positive ($\partial^2 \Delta G/\partial x^2 > 0$).

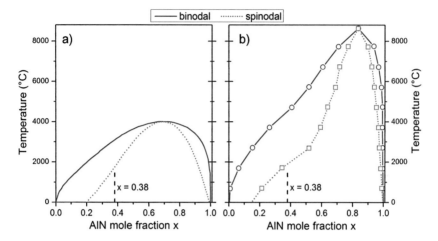

Fig. 7.11: Temperature phase diagram for fcc-Ti$_{1-x}$Al$_x$N calculated by Mayrhofer *et al.* [204] (a) and Alling *et al.* [205] (b) showing the binodal and spinodal. Each diagram is reproduced after Ref. [204] and Ref. [205], respectively. The Ti$_{1-x}$Al$_x$N coatings with x = 0.38 lie within the spinodal region as indicated by the dashed black line.

7.1.7.1 Thermally activated microstructure changes in the Ti$_{0.62}$Al$_{0.38}$N coatings

Before annealing the **Ti$_{0.62}$Al$_{0.38}$N coating deposited at U$_B$ = -40 V** contained fcc-(Ti,Al)N as single phase, as only the diffraction lines of fcc-(Ti,Al)N of the coating and the diffraction lines of fcc-TiC and hex-WC of the substrate were visible in the diffraction pattern shown in Fig. 7.12a. The single-phase nature of the as-deposited sample was confirmed by the stress-free lattice parameter because it agrees quite well with the stress-free lattice parameter expected for the coating composition (see lower dashed line in Fig. 7.13b) according to the Vegard-like dependence (Eq. (7.1)). Due to the agreement between the measured and expected stress-free lattice parameter, it was concluded that the whole Al present in the coating is incorporated into fcc-(Ti,Al)N. The as-deposited coating is characterized by low macroscopic lattice strain $-(1.0 \pm 0.4) \times 10^{-3}$ that indicates a low compressive stress, which is shown additionally in Fig. 7.13c and as calculated for $E = 500$ GPa and $\nu = 0.3$. The stress-free lattice parameter increases at 450°C and 650°C just due to thermal expansion of fcc-(Ti,Al)N because cooling the coating to 100°C yielded the same a_0 as in the initial state. No macroscopic lattice strain or residual stress was observed when the temperature decreased from 450°C to 100°C.

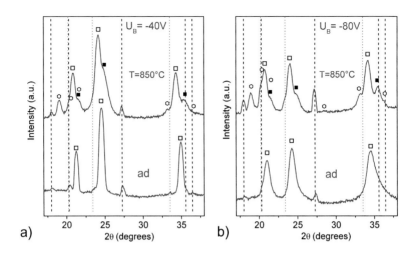

Fig. 7.12: Parts of the diffraction patterns obtained from the *in situ* GAXRD measurements of the as-deposited state and at 850°C (after ~ 150 min annealing) of the Ti$_{0.62}$Al$_{0.38}$N coating deposited at U$_B$ = -40 V (a) and U$_B$ = -80 V (b). The intensities are plotted in logarithmic scale. The symbols indicate the positions of fcc-(Ti,Al)N (□), fcc-AlN (■) and w-AlN (hexagon) diffraction lines. The positions of hex-WC and fcc-TiC diffraction lines from the substrate are marked by the dashed and dotted lines, respectively.

Annealing the coating to 850°C led to significant changes in the microstructure. Already in the first HT-GAXRD pattern at 850°C, which was obtained after ~ 40 min at this temperature, diffraction lines of w-AlN and fcc-AlN became visible in the HT-GAXRD pattern indicating the decomposition of fcc-(Ti,Al)N. Their intensity increased during further annealing at this temperature as shown in Fig. 7.12a. The formation of w-AlN already at 850°C was unexpected since the formation of w-AlN was reported frequently to take place after annealing at temperatures above 1000°C e.g. [44, 73, 74] (see also Section 3.3). However, recent publications support the observation that w-AlN can already be formed between 850 - 900°C [80, 81, 84, 94]. In the Ti$_{0.62}$Al$_{0.38}$N coating deposited at U$_B$ = -40 V both AlN modifications exist simultaneously at 850°C. The mole fraction of w-AlN is below 10 % whereas the mole fraction of fcc-AlN is approx. four times higher (see Fig. 7.13a). Due to the formation of both AlN phases the [Al] / ([Ti]+[Ti])-ratio in the fcc-(Ti,Al)N phase decreased. As a result of the decomposition of fcc-(Ti,Al)N and due to thermal expansion, the stress-free lattice parameter of the fcc-(Ti,Al)N phase increased strongly at 850°C even above the value of Al-free fcc-TiN corresponding to room temperature ($a_{fcc-TiN}$ = 0.4242 nm) (see Fig. 7.13b). At the beginning

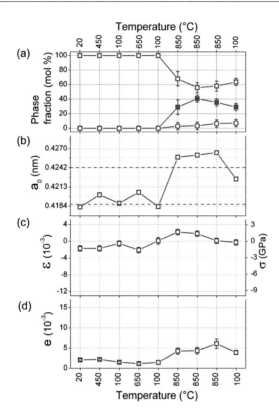

Fig. 7.13: Evolution of microstructure parameters in the Ti$_{0.62}$Al$_{0.38}$N coating deposited at U$_B$ = -40 V during thermal treatment showing the change of the phase fractions of fcc-(Ti,Al)N ($\square$), fcc-AlN ($\blacksquare$) and w-AlN (hexagon) (a) as estimated from Rietveld analysis and the change of the stress-free lattice parameter (b), macroscopic lattice strain and residual stress (c) and apparent microstrain (d) of the fcc-(Ti,Al)N phase. The dashed red lines in (b) correspond to a$_0$ of fcc-TiN (0.4242 nm) and fcc-Ti$_{0.62}$Al$_{0.38}$N (0.4187 nm) as calculated from the Vegard-like dependence at room temperature (Eq. (7.1)).

of the phase decomposition at 850°C tensile stress was observed in the fcc-(Ti,Al)N phase (see Fig. 7.13c). According to Ref. [82] this effect could be attributed to the growth of fcc-(Ti,Al)N/fcc-AlN interfaces that grow perpendicular to the sample surface. Due to the smaller lattice parameter of fcc-AlN, fcc-(Ti,Al)N is compressed at the interface to fcc-AlN. This appears like tensile residual stress in the sin$^2 \psi$ method since the maximum compression is perpendicular to the sample surface. The apparent microstrain increased in the fcc-(Ti,Al)N phase (see Fig. 7.13d) due to the rearrangement of the Al atoms taking place during decomposition. The analysis of the stress-free lattice parameter after annealing for ~ 150 min at 850°C min revealed that the Al content in the fcc-Ti$_{1-x}$Al$_x$N phase was reduced to x = 0.12. The reduction of the Al content in the fcc-Ti$_{1-x}$Al$_x$N confirms the observed segregation of AlN in order to form w-AlN and fcc-AlN. Since the determined lattice parameters of the wurtzitic phase were a = 0.309 nm and c = 0.498 nm and approached the expected ones [190], no remnants of Ti in the wurtzitic phase is suspected after annealing at 850°C. The phase composition as estimated by Rietveld analysis revealed that beside the major fcc-Ti$_{0.88}$Al$_{0.12}$N phase a considerable amount of ~ 29 mol% fcc-AlN could be stabilized in the coating, whereas just 7 mol% w-AlN were formed.

The development of the microstructure parameters of the **Ti$_{0.62}$Al$_{0.38}$N coating deposited at U$_B$ = -80 V** represents the second type of the decomposition kinetics. The stress-free lattice parameter of the fcc-(Ti,Al)N phase in the initial state was larger than the lattice parameter predicted from the Vegard-like dependence according to Eq. (7.1) for the Ti$_{0.62}$Al$_{0.38}$N coating (see Fig. 7.14b). The experimentally determined a_0 corresponds to fcc-Ti$_{0.73}$Al$_{0.27}$N. This means that not the whole Al in the coating is incorporated into the fcc-Ti$_{1-x}$Al$_x$N phase but forms instead another Al-rich phase or is present in an amorphous phase. Due to the asymmetry of the diffraction lines of the fcc-(Ti,Al)N phase as shown in Fig. 7.12b and discussed in Section 7.1.2, it was concluded that minor fractions of fcc-AlN or Al-rich fcc-(Al,Ti)N are present. Since no distinct peaks of fcc-AlN or w-AlN were visible in the GAXRD pattern, Rietveld analysis yielded 100 mol% fcc-(Ti,Al)N (see open squares in Fig. 7.14a). Under the assumption that the Al which is not incorporated into the fcc-Ti$_{0.73}$Al$_{0.27}$N phase forms crystalline fcc-AlN, the phase composition would yield 85 mol% fcc-Ti$_{0.73}$Al$_{0.27}$N and 15 mol% fcc-AlN in the coating with the overall composition Ti$_{0.62}$Al$_{0.38}$N. The phase fraction of fcc-AlN that was recalculated from the a_0 of the fcc-(Ti,Al)N phase that was measured in the initial state and at 100°C is plotted by triangles in Fig. 7.14a. The dual phase nature of the as-deposited coating contributed to large macroscopic lattice strain (see Fig. 7.14c). After the annealing steps at 450°C and 650°C the lattice parameter of fcc-(Ti,Al)N that was measured at 100°C decreased gradually. After cooling the sample from 650°C to 100°C, the stress-free lattice parameter matched to the coating's overall composition according to the Vegard-like dependence (Eq. (7.1)). Furthermore, the asymmetry of the diffraction lines of fcc-(Ti,Al)N decreased. This indicated the complete incorporation of AlN into fcc-(Ti,Al)N after such a thermal cycle although the incorporation of AlN into fcc-TiN would not be expected from the thermal equilibrium of this system [42]. This process was accompanied by the reduction of the compressive stress (see Fig. 7.14c). Apart from the incorporation of AlN into fcc-(Ti,Al)N, the annihilation of point defects that were induced by ion impact during the deposition process could be considered as a reason for the reduction of a_0 and the compressive stress [66]. However, implantation experiments of TiN with Al ions [206] and N^+ [207] having an energy of 60 - 70 keV and a density of 10^{17} ions/cm^2 and 3×10^{17} ions/cm^2, respectively, indicated only a small contribution of point defects to the change of the lattice parameter. The generated point defects in References [206, 207] caused a relative increase of the near surface stress-free lattice parameter of 10^{-3} that is substantially smaller than the relative decrease in the stress-free lattice parameter of 6×10^{-3} in the annealed Ti-Al-N coating of this study.

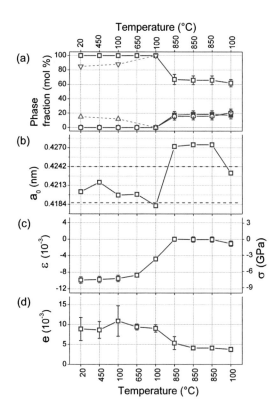

Fig. 7.14: Evolution of microstructure parameters in the Ti$_{0.62}$Al$_{0.38}$N coating deposited at $U_B = -80$ V during thermal treatment showing the change of the phase fractions (a) as estimated from Rietveld analysis and the change of the stress-free lattice parameter (b), macroscopic lattice strain and residual stress (c) and apparent microstrain (d) of the fcc-(Ti,Al)N phase. The green triangles indicate the content of fcc-(Ti,Al)N (∇) and fcc-AlN ($\triangle$) as calculated from a_0. The symbols and lines have the same meaning like in Fig. 7.13.

During annealing at 850°C, AlN segregated from fcc-(Ti,Al)N. Diffraction lines of w-AlN and fcc-AlN became clearly visible in the HT-GAXRD pattern as shown in Fig. 7.12b. Already after 40 min during annealing at 850°C similar mole fractions of w-AlN and fcc-AlN of ~ 20 mol% appeared in the coating. The phase fractions did not change significantly with proceeding annealing time at 850°C. The decomposition process at 850°C was accompanied by the reduction of the compressive stress (Fig. 7.14c) and decrease of the microstrain (Fig. 7.14d). After cooling the sample to 100°C, the stress-free lattice parameter of the Ti-rich fcc-$Ti_{1-x}Al_xN$ phase indicated an Al content of $x < 0.1$ in it.

The microstructure evolution of the **$Ti_{0.62}Al_{0.38}N$ coating deposited at $U_B = -120$ V** represents the third type of the decomposition kinetics. Similar to the $Ti_{0.62}Al_{0.38}N$ coating deposited at $U_B = -80$ V, only diffraction lines of fcc-(Ti,Al)N but no distinct diffraction lines of w-AlN and fcc-AlN were present in the GAXRD pattern of the as-deposited coating (see Fig. 7.2d). Thus, the phase fraction of fcc-(Ti,Al)N estimated by Rietveld analysis yielded 100 mol% (see open squares in Fig. 7.15a). But fitting of the GAXRD pattern recognized a slight asymmetry of the fcc-(Ti,Al)N diffraction lines particularly for the diffraction lines with even *hkl* (see e.g. Fig. 7.2d). Additionally, the stress-free lattice parameter of fcc-(Ti,Al)N was larger as expected for fcc-$Ti_{0.62}Al_{0.38}N$ (Fig. 7.15b). These findings led to the conclusion, that fcc-AlN is present as second phase in this coating. The re-calculation of the measured a_0 of the fcc-(Ti,Al)N phase into its chemical composition and the comparison with the overall chemical composition of the coating revealed ~ 93 mol% fcc-$Ti_{0.67}Al_{0.33}N$ and 7 mol% fcc-AlN, which is shown by open triangles in Fig. 7.15a. Like for the $Ti_{0.62}Al_{0.38}N$ coating deposited at $U_B = -80$ V, the dual-phase nature of the coating deposited at $U_B = -120$ V is accompanied by high macroscopic lattice strains indicating a high compressive stress state (Fig. 7.15c). Annealing at 450°C did not cause a sustainable change of the microstructure, since the observed microstructure parameters in Fig. 7.15 behaved reversibly upon cooling. The situation changed during annealing at 650°C, because the GAXRD pattern obtained after cooling from 650 to 100°C revealed the incorporation of AlN into fcc-(Ti,Al)N as concluded from the decrease of the stress-free lattice parameter of fcc-(Ti,Al)N to the value expected for the coating composition (lower dashed line in Fig. 7.15b). In contrast to the $Ti_{0.62}Al_{0.38}N$ coatings deposited at $U_B = -40$ V and $U_B = -80$ V the compressive stress of the fcc-(Ti,Al)N phase was reduced only slightly after the annealing step at 650°C as shown in Fig. 7.15c. First during annealing at 850°C the macroscopic lattice strain and compressive stress relaxed. Already within 40 min during annealing at 850°C a considerable amount of w-AlN was formed, whereas the phase fraction of fcc-AlN was the lowest one among the $Ti_{0.62}Al_{0.38}N$ coatings deposited at different bias voltages (see Fig. 7.15a).

Similar to the $Ti_{0.62}Al_{0.38}N$ coatings deposited at $U_B = -80$ V, the apparent microstrain in the fcc-(Ti,Al)N phase decreased significantly during decomposition at 850°C (Fig. 7.15d). The analysis of the stress-free lattice parameter after cooling from 850°C to 100°C indicated that nearly the whole Al left the fcc-(Ti,Al)N host structure and formed AlN.

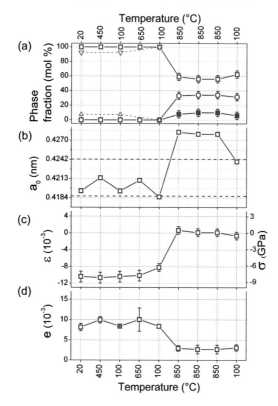

Fig. 7.15: Evolution of microstructure parameters in the $Ti_{0.62}Al_{0.38}N$ coating deposited at $U_B = -120$ V during thermal treatment showing the change of the phase fractions (a) as estimated from Rietveld analysis and the change of the stress-free lattice parameter (b), macroscopic lattice strain and residual stress (c) and apparent microstrain (d) of the fcc-(Ti,Al)N phase. The symbols and lines have the same meaning like in Fig. 7.13 and Fig. 7.14.

7.1.7.2 Effect of the bias voltage and chemical composition on the phase composition of annealed Ti₁₋ₓAlₓN coatings

A similar correlation of the bias voltage, that was applied during the deposition of $Ti_{1-x}Al_xN$ coatings, and the thermally activated evolution of the microstructure as described above for the $Ti_{0.62}Al_{0.38}N$ coatings was also observed for other $Ti_{1-x}Al_xN$ coatings with $x = 0.47$ and $x = 0.56$. Especially the incorporation of AlN into the host structure of fcc-(Ti,Al)N at 650°C, the decomposition of fcc-(Ti,Al)N into w-AlN/fcc-AlN and Al-depleted fcc-(Ti,Al)N at 850°C, the almost entire relaxation of the compressive stress during annealing at 850°C and the reduction of the apparent microstrain for the coatings deposited at $U_B = -80$ V and $U_B = -120$ V were observed in a slightly modified form for the other investigated chemical compositions of the $Ti_{1-x}Al_xN$ coatings. In order to summarize the effect of the chemical composition and the bias voltage on the thermal stability of the investigated $Ti_{1-x}Al_xN$ coatings Fig. 7.16 was chosen. It illustrates the phase composition in the as-deposited state (Fig. 7.16a) and after annealing at 850°C (Fig. 7.16b) as well as the apparent microstrain before and after the thermal treatment (Fig. 7.16c). It is obvious that the bias voltage not only influences the phase composition in the as-deposited, as already discussed in Section 7.1.2, but also the phase composition after annealing at 850 °C. In the as-deposited coatings a high bias voltage (-80 V and -120 V) led to

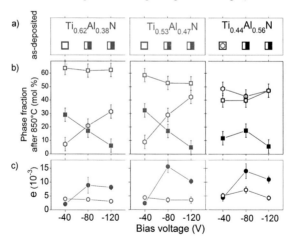

Fig. 7.16: Summary of the phase composition in the as-deposited coatings (a) and after annealing at 850 °C estimated from Rietveld analysis (b) as well as the apparent microstrain in the as-deposited state (•) and after 850 °C (○) (c) for the investigated Ti₁₋ₓAlₓN coatings. The symbols in (a) denote the present phases namely fcc-(Ti,Al)N (□), fcc-(Ti,Al)N/fcc-AlN (▣) and w-AlN (hexagon). The symbols in (b) have the same meaning like in Fig. 7.13. The lines are guide for the eyes.

the formation of fcc-(Ti,Al)N and minor amounts of fcc-AlN. The annealing experiments revealed also a correlation between the bias voltage and the relative amount of w-AlN and fcc-AlN in the $Ti_{0.62}Al_{0.38}N$, $Ti_{0.53}Al_{0.478}N$ and $Ti_{0.44}Al_{0.56}N$ coatings after annealing at 850°C. The molar fraction of w-AlN increased at the expense of fcc-AlN with increasing bias voltage. Only the $Ti_{0.44}Al_{0.56}N$ coating deposited at $U_B = -40$ V deviated from this trend. In this coating w-AlN was present already in the as-deposited state and after annealing at 850°C it represented the highest mole fraction of w-AlN whereas the fraction of fcc-AN was low. The influence of the bias voltage on the molar fraction of fcc-(Ti,Al)N after annealing at 850°C is low. Its amount is predominantly controlled by the overall chemical composition of the coatings.

The results shown in Sections 7.1.7.1 and 7.1.7.2 suggest that the bias voltage, which adjusted the microstructure of the as-deposited coatings, determines the decomposition kinetics of the $Ti_{1-x}Al_xN$ coatings during annealing. The increase of the phase fraction of w-AlN and decrease of fcc-AlN with increasing bias voltage (see Fig. 7.16b) that was found after annealing at 850°C suggested an accelerated decomposition process in the coatings that were deposited at high bias voltages. The kinetics of the decomposition seems to be determined by the different reduction of the macroscopic lattice strain that was observed for the $Ti_{1-x}Al_xN$ coatings deposited at $U_B = -40$ V, $U_B = -80$ V and $U_B = -120$ V and shown in detail for the $Ti_{0.62}Al_{0.38}N$ coatings (see Fig. 7.13c, Fig. 7.14c and Fig. 7.15c). The low macroscopic lattice strain in the single-phase coatings relaxed already when the coating was cooled from 450°C to 100°C. The reduction of the compressive stress in the dual-phase coatings started after the annealing at 650°C. At the same time the equalization of local compositional fluctuations was observed that led to the incorporation of AlN into the fcc-(Ti,Al)N host structure. This process seemed to contribute to the reduction of the large compressive stress. As shown exemplarily for the $Ti_{0.62}Al_{0.38}N$ coatings, the reduction of the compressive stress before the 850°C annealing step was retarded in the coating deposited at $U_B = -120$ V as compared to $U_B = -80$ V. The different reduction of the lattice strains seems to influence the kinetics of the phase decomposition into fcc-AlN, w-AlN and Al-depleted fcc-(Ti,Al)N. The effect of lattice strains on the decomposition was already proposed by Cahn [208] and indicated in the calculations and experiments in Refs. [81, 209]. Within the experiments performed in this study, the retarted reduction of the macroscopic lattice strain seemed to accelerate the formation of w-AlN as it was observed for the coatings deposited at $U_B = -120$ V. A similar effect of the stress was found by Schalk et al. [81] in the course of a thermal treatment of magnetron sputtered Ti-Al-N coatings.

7.1.7.3 Apparent microstrain as an indicator of local concentration fluctuations in fcc-(Ti,Al)N

Rachbauer *et al.* [195] demonstrated by atom probe analysis of an as-deposited $Ti_{0.46}Al_{0.54}N$ coating, which was prepared by magnetron sputtering, that the Al content is not constant through the fcc-(Ti,Al)N phase it rather fluctuates around its mean value. These local variations of the [Al] / ([Al]+[Ti]) ratio on the nm-scale are accompanied by local changes of the interplanar spacing $(\Delta d / d)$.

In general the relative change of the interplanar spacing (see Eq. (7.11)) can be obtained from the differential form of the Bragg equation for a constant wavelength of the X-rays (see Eq. (7.10)):

$$2\Delta d \sin\theta + 2d \,\Delta\theta\cos\theta = 0 \tag{7.10}$$

$$-\frac{\Delta d}{d} = \Delta\theta \cot\theta \tag{7.11}$$

where θ is one half of the diffraction angle (2θ) and $\Delta\theta$ corresponds to the line broadening caused by the variations of the interplanar spacing. The root mean square of the variations of the interplanar spacings corresponds to the microstrain (e):

$$e = \left\langle \left(\frac{\Delta d}{d}\right)^2 \right\rangle^{\frac{1}{2}} \tag{7.12}$$

The microstrain can be determined from the slope (m_{WH}) of the increasing line broadening (given in inverse unit of length) with increasing $\sin\theta$ in the Williamson-Hall plot [210] according to Eq. (7.13):

$$e = \frac{m_{WH} \cdot \lambda}{4} \tag{7.13}$$

The line broadening is also often plotted versus the modulus of the diffraction vector which corresponds to:

$$|\vec{q}| = q = \frac{4\pi \sin\theta}{\lambda} \tag{7.14}$$

In general, several reasons for the increase of the line broadening with rising q are possible. One reason can be the inhomogeneous variations of the interplanar spacing which causes the lattice strains as shown by Eq. (7.12). The inhomogeneous variations of the interplanar spacing can be caused by dislocations, dislocation structures, lattice misfit at the interfaces between fcc-(Ti,Al)N and w-AlN as well as non-uniformly distributed aluminium atoms.

Another reason for the increase of the line broadening with increasing q is the decay of the partial coherence with increasing q [200]. In case of partially coherent nanocrystallites, the reciprocal lattice points of the crystallites overlap at low diffraction angles leading to a small line width of the diffraction lines. With increasing diffraction vector, the overlap of the reciprocal lattice decreases and the line width increases. The slope of the line broadening vs. the modulus of the diffraction vector is determined by the number of coherent crystallites. Many simultaneously partial coherent crystallites cause a lower increase of the line broadening with increasing q [82].

Due to the possible origins of the line broadening a direct identification of the reasons for the line broadening is difficult. Therefore, the microstrain determined from the observed line broadening of the investigated coatings is denoted as *apparent microstrain*.

If this apparent microstrain would be attributed solely to the change of the interplanar spacing caused by the local variations of the [Al] / ([Al]+[Ti]) ratio, the local variations of the Al content in the fcc-(Ti,Al)N phase could be calculated from the apparent microstrain.

In cubic materials, the relative change of the interplanar spacing is equal to the relative change of the lattice parameter:

$$\frac{\Delta d}{d_0} = \frac{\Delta a}{a_0} \qquad (7.15)$$

The change in the lattice parameter Δa of fcc-$Ti_{1-x}Al_xN$ due to the change of the Al concentration Δx can be described by the differential form of the Vegard-like dependence represented by Eq. (7.1):

$$\Delta a = -0.01432 \text{ nm} \cdot \Delta x \qquad (7.16)$$

Thus the relative change of the lattice parameter is given by Eq. (7.17). The negative value of the coefficient in Eq. (7.17) indicates that the increase of the Al content causes a decrease of the lattice parameter of the fcc-$Ti_{1-x}Al_xN$ phase.

$$\frac{\Delta a}{a_0} = -0.01432 \text{ nm} \cdot \frac{\Delta x}{a_0} \qquad (7.17)$$

The combination of Eq. (7.12) with Eqs. (7.15) and (7.17) yields the relationship between the apparent microstrain (e), the mean stress-free lattice parameter (a_0) and the root-mean-squared change in the Al concentration $\langle(\Delta x)^2\rangle^{\frac{1}{2}}$:

$$\langle(\Delta x)^2\rangle^{\frac{1}{2}} = \frac{a_0 \cdot e}{0.01432 \text{ nm}} \qquad (7.18)$$

According to Eq. (7.18) the composition of the fcc-(Ti,Al)N phase would vary between $Ti_{1-(x+\Delta x/2)}Al_{(x+\Delta x/2)}N$ and $Ti_{1-(x-\Delta x/2)}Al_{(x-\Delta x/2)}N$. Assuming that the slope of the line broadening in the Williamson-Hall plot is solely influenced by the concentration fluctuations, the fluctuation of the [Ti] / ([Ti]+[Al]) ratio in the fcc-(Ti,Al)N phase was calculated from the apparent line broadening using Eq. (7.18) and is shown in Fig. 7.17 for the as-deposited state and after annealing at 850°C.

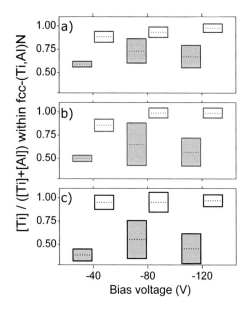

Fig. 7.17: Mean Ti contents in the fcc-(Ti,Al)N phase and their fluctuations as a function of the bias voltage for the $Ti_{0.62}Al_{0.38}N$ (a), $Ti_{0.53}Al_{0.47}N$ (b) and $Ti_{0.44}Al_{0.56}N$ (c) coatings. The grey and open boxes correspond to the as-deposited states and to the states after annealing at 850°C, respectively.

According to this simplified model, very shallow concentration fluctuations of the metallic species within fcc-(Ti,Al)N are present in the as-deposited single-phase $Ti_{0.62}Al_{0.38}N$ and $Ti_{0.53}Al_{0.47}N$ coatings that were produced at U_B = -40 V (see grey boxes in Fig. 7.17a and Fig. 7.17b). The annealing at 850 °C of these coatings led to the out-diffusion of AlN and the formation of fcc-AlN and w-AlN. Consequently, after the completion of the thermal cycle the [Ti] / ([Ti]+[Al]) ratio in the fcc-(Ti,Al)N phase increased. Concurrently, the local concentration fluctuations increased slightly due to the rearrangement of the Al atoms taking place during decomposition, as illustrated by the height of the open boxes in Fig. 7.17.

The coatings which were deposited at U_B = -80 V and U_B = -120 V were already dual-phase in the as-deposited state since the presence of fcc-AlN was supposed beside the major phase fcc-(Ti,Al)N (see Section 7.1.2). Due to that, the [Ti] / ([Ti]+[Al]) ratio in the fcc-(Ti,Al)N phase is larger than in the single-phase coatings with the same chemical composition, as shown by the black dotted line in Fig. 7.17. The highest mean $[Ti] / ([Ti]+[Al])$ ratio is found in the $Ti_{1-x}Al_xN$ coatings deposited at $U_B = -80$ V among the as-deposited coatings with the same chemical composition. Furthermore, the dual-phase nature in the as-deposited coatings is connected with a larger amplitude of local concentration fluctuations as compared to the single-phase coatings. However, the apparent microstrain indicates the highest concentration fluctuations in the as-deposited coatings that were produced at $U_B = -80$ V. The out diffusion of AlN from the fcc-(Ti,Al)N phase, as a result of the thermal treatment at 850°C, led to a significant lower amplitude of local concentration fluctuations in the fcc-(Ti,Al)N phase and an increased $[Ti] / ([Ti]+[Al])$ ratio in the fcc-(Ti,Al)N phase in the $Ti_{1-x}Al_xN$ coatings deposited at $U_B = -80$ V and $U_B = -120$ V. However, the limitations of the method that was used to estimate the concentration fluctuations from the apparent microstrain become visible especially in Fig. 7.17b and Fig. 7.17c for the $Ti_{1-x}Al_xN$ coatings deposited at $U_B = -80$ V and $U_B = -120$ V after the completion of the thermal cycle. In these cases, the $[Ti] / ([Ti]+[Al])$ ratio exceeds slightly the maximum Ti content in fcc-(Ti,Al)N which is equal to $[Ti] / ([Ti]+[Al]) = 1$. This indicates, that the line broadening is not solely caused by the local concentration fluctuations.

7.2 Ti-Al-N / Al-Ti-(Ru)-N multilayers deposited by CAE

The deposition of multilayers can be used as complementary method to design nanocomposites [131]. Using a multilayer architecture, materials with different functional characteristics can be combined. Furthermore, the resistance of crack propagation can be increased by the presence of interfaces [211] or by a compositional modulation [2]. A special type of multilayers represent the so called "superlattices" with a bi-layer period of typically ~ 4 nm [212] that are characterized by a hardness increase by a factor of 2 to 7 as compared to the individual layer materials [127].

In industrial applications, multilayers based on Ti-Al-N are appreciated as protective layers especially due to their improved thermal stability, oxidation resistance and better wear properties as compared to monolithic coatings. For example, the combination of an Al-rich (Al,Ti)N layer with an fcc-TiN layer has be shown to provide a better thermal stability than the monolithic coatings [40].

By the addition of doping elements to the Ti-Al-N based multilayer coatings further improvements of the coating properties are aspired. In Ref. [101] it was shown that doping of Ti-Al-N / Al-Ti-N multilayer coatings with Ru improved the lifetime of the coated cutting inserts. Thus, in the frame of this work the effect of the Ru addition and the bias voltage, which modified the energetic input of the ions to the growing film, on the microstructure, hardness and thermal stability of Ti-Al-N / Al-Ti-(Ru)-N multilayers was investigated and is shown in this Section. The results were published partially in Ref. [213, 214].

7.2.1 Deposition of CAE Ti-Al-N / Al-Ti-(Ru)-N multilayer coatings

The Ti-Al-N / Al-Ti-(Ru)-N multilayer coatings were deposited in the same CAE facility of the Balzers RCS type like the Ti-Al-N coatings (see Section 7.1.1). The multilayers were deposited

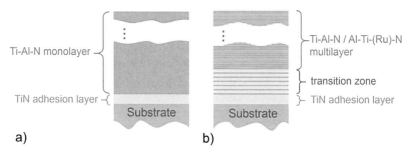

Fig. 7.18: Sketch of the coating architecture of a Ti-Al-N monolayer coating (a) and a Ti-Al-N / Al-Ti-(Ru)-N multilayer coating (b). The sketch is not true to scale.

from two $Ti_{50}Al_{50}$ and two $Ti_{33-y}Al_{67}Ru_y$ PM targets from PLANSEE CM [186]. The $Ti_{50}Al_{50}$ targets were mounted on the positions 5 and 6 (see Fig. 7.1) and the $Ti_{33-y}Al_{67}Ru_y$ targets were attached to the positions 1 and 4 in the deposition chamber. During deposition a two-fold rotation was applied. The rotation speed of the carousel was set to 37 s per rotation. Three coating series with different Ru content were deposited from $Ti_{33-y}Al_{67}Ru_y$ targets with different Ru content: (I) $y = 0$, (II) $y = 1$ and (III) $y = 5$. Each series was deposited at four different bias voltages of -20 V, -40 V, -60 V and -80 V in order to modify the ion impact during deposition. The temperature, pressure of the nitrogen atmosphere and the deposition time was the same like for the $Ti_{1-x}Al_xN$ monolayer coatings (see Section 7.1.1). Before the deposition of the actual Ti-Al-N / Al-Ti-(Ru)-N multilayer a TiN adhesion layer was deposited onto the cleaned substrates by the evaporation of titanium PM targets which were mounted on the positions 2 and 3 (see Fig. 7.1). After the growth of a ~ 0.2 μm thick fcc-TiN adhesion layer, a transition zone consisting of six stacks (see sketch in Fig. 7.18b), that were composed of ~ 12 nm thick Ti-Al-(Ru)-N and ~ 25 nm thick TiN layers, were deposited due to the carousel rotation and the simultaneous operation of the Ti and $Ti_{33-y}Al_{67}Ru_y$ targets.

Finally, after the deposition of the ~ 0.22 μm thick transition zone, the Ti targets were switched off and the actual Ti-Al-N / Al-Ti-(Ru)-N multilayers were deposited from the $Ti_{33-y}Al_{67}Ru_y$ and $Ti_{50}Al_{50}$ targets. Apart from three multilayer coatings of series I that were deposited at $U_B = -20$ V, -60 V and -80 V, the same type of cemented carbides like for the Ti-Al-N monolayer coatings was used. The three exceptions ($U_B = -20$ V, -60 V and -80 V) from coating series I were deposited onto cemented carbide substrates containing WC with 10 wt.% Co as well as on Si (100) substrates with the dimension of $25 \times 15 \times 0.6$ mm^3.

7.2.2   Chemical and phase composition of as-deposited Ti-Al-N / Al-Ti-(Ru)-N multilayers

The overall metal ratios within the coatings were analyzed by EPMA / WDS and are shown in Table 7.1. The results were confirmed by GDOES measurements. The concentration fluctuations attributed to the multilayer architecture could not be resolved by GDOES. The overall $[Al] / ([Al] + [Ti] + [Ru])$ ratio (abbreviated as $[A] / \Sigma[Me]$ in the following) of the targets corresponds to 0.585 for all coating series. However, the analysis of the chemical composition of the coatings revealed a $[A] / \Sigma[Me]$ ratio between 0.52 and 0.53 in all coatings which is about 10 % lower as compared to the average $[A] / \Sigma[Me]$ ratio of the targets. This effect was already reported in the literature for coatings deposited by CAE and is attributed to an increased ionization of titanium (see also Section 3.2.2.1) and preferential re-sputtering of already deposited Al atoms [60, 65]. Consequently, the average $[Al] / ([Al] + [Ti] + [Ru])$ ratio (abbreviated as $[Al] / \Sigma[Me]$) in the coatings is higher than in the targets. Furthermore, the $[Ti] / \Sigma[Me]$ ratio decreases from coating series I to III due to the increasing Ru content at the expense of the Ti content in the targets. No correlation between the bias voltage and the preferential incorporation of Al, Ti or Ru into the coatings was observed.

Table 7.1: Metal ratios of the Ti-Al-N / Al-Ti-Ru-N multilayers as determined by using EPMA / WDS; $\Sigma[Me] = ([Ti] + [Al] + [Ru])$.

Coating series	Target combination	$\dfrac{[Ti]}{\Sigma[Me]}$	$\dfrac{[Al]}{\Sigma[Me]}$	$\dfrac{[Ru]}{\Sigma[Me]}$
I	$2\times$ Ti$_{50}$Al$_{50}$ & $2\times$ Ti$_{33}$Al$_{67}$	0.48 ± 0.01	0.52 ± 0.01	0
II	$2\times$ Ti$_{50}$Al$_{50}$ & $2\times$ Ti$_{32}$Al$_{67}$Ru$_{1}$	0.47 ± 0.01	0.53 ± 0.01	$0.005 \pm < 0.001$
III	$2\times$ Ti$_{50}$Al$_{50}$ & $2\times$ Ti$_{28}$Al$_{67}$Ru$_{5}$	0.45 ± 0.01	0.53 ± 0.01	$0.02 \pm < 0.001$

The phase composition of the coatings was investigated qualitatively by GAXRD. Parts of the GAXRD patterns of coating series II and III are shown in Fig. 7.19. The diffraction patterns of series I were similar. The GAXRD patterns showed the diffraction lines of fcc-(Ti,Al)N in all coatings. No distinct diffraction lines of w-AlN were visible in the GAXRD patterns, but the increased intensity between $32°$ and $35.2°$ 2θ indicates the presence of traces of w-AlN in all coatings. The assumption that traces of w-AlN are present in the coatings is supported by the GAXRD patterns obtained with synchrotron radiation ($\lambda = 0.10781$ nm). Due to the high flux of the beam, the presence of the low scattering w-AlN phase in the coatings becomes more apparent in the diffraction patterns as compared to laboratory experiments with Cu radiation. This is shown exemplarily for the coatings of series III that were deposited at $U_B = -40$ V, $U_B = -60$ V and $U_B = -80$ V (see Fig. 7.20).

Furthermore, the GAXRD patterns obtained by laboratory experiments (see Fig. 7.19) as well as by synchrotron experiments (see Fig. 7.20) showed an asymmetric peak broadening to higher diffraction angles of the fcc-(Ti,Al)N diffraction lines in all coatings that were deposited at high bias voltages of $U_B = -60$ V and especially at $U_B = -80$ V. This effect was also observed in Ti$_{1-x}$Al$_x$N monolayer coatings (see Section 7.1.2). Due to the asymmetric peak broadening, the fcc-(Ti,Al)N diffraction lines could only be described by two symmetrical Pearson VII functions. The first peak belongs to the main fcc-(Ti,Al)N phase. The second peak corresponds to a weak and broadened diffraction line of an fcc-phase with lower lattice parameter which could be attributed to fcc-AlN or an Al-rich fcc-(Ti,Al)N phase [93].

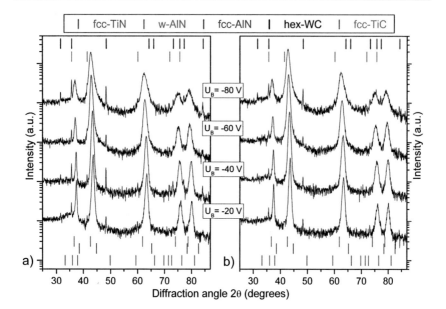

Fig. 7.19: Parts of the GAXRD patterns of the Ti-Al-N / Ti-Al-Ru-N multilayer coatings from series II (a) and III (b) deposited at different bias voltages as measured by Cu radiation. The positions of the w-AlN, fcc-AlN and fcc-TiN diffraction lines are labelled at the bottom of the figure and the positions of the fcc-TiC and hex-WC diffraction lines coming from the substrate are shown at the top of the figure.

Further information about the phase composition of the coatings could be deduced from the volume-weighted stress-free lattice parameter of the fcc-(Ti,Al)N phase [82] which was evaluated from the GAXRD patterns. In order to obtain a_0, at first a_ψ^{hkl} was plotted vs. $\sin^2 \psi$ as shown in Fig. 7.21a. The $\sin^2 \psi$-plot revealed a large scatter of the lattice parameter a_ψ^{hkl} vs. $\sin^2 \psi$. This indicated the presence of crystal anisotropy of the lattice deformation of the fcc-phase in the Ti-Al-N / Al-Ti-(Ru)-N multilayer coatings. The measurement of the interplanar spacing of the lattice planes (111) and (200) at different sample inclinations, as shown by blue open circles and boxes respectively for the Ti-Al-N / Al-Ti-Ru-N multilayer coating of series II deposited at $U_B = -60$ V, illustrated the anisotropy of the elastic lattice strain with $\varepsilon^{h00}/\varepsilon^{hhh} > 1$, which means that the easy deformation direction is $\langle 100 \rangle$ and the hard deformation direction is $\langle 111 \rangle$ in the investigated coatings. This corresponds to an anisotropy factor $A > 1$ (see Eq. (6.41)) which is in agreement with Ref. [215] that predicted an anisotropy factor $A > 1$ for $x > 0.28$ in $Ti_{1-x}Al_xN$ coatings using *ab initio* calculations.

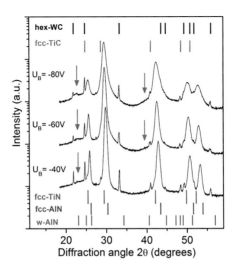

Fig. 7.20: Low angle parts of the GAXRD patterns of the Ti-Al-N / Ti-Al-Ru-N multilayer coatings of series III deposited at $U_B = -40$ V, $U_B = -60$ V and $U_B = -80$ V as measured using synchrotron radiation ($\lambda = 0.10781$ nm). The positions of the w-AlN, fcc-AlN and fcc-TiN diffraction lines are labelled at the bottom of the figure and the positions of the fcc-TiC and hex-WC diffraction lines coming from the substrate are shown at the top of the figure. The green arrows indicate the increased intensity in the diffraction patterns caused by traces of wurtzite.

In order to determine a_0 of the fcc-(Ti,Al)N phase in the presence of crystal anisotropy of the lattice deformation of the fcc-phase, the dependence of the lattice parameter a_ψ^{hkl} on $\sin^2 \psi$ and on the orientation factor Γ was described by Eq. (6.27) at first. For the example of the Ti-Al-N / Al-Ti-Ru-N coating of series II deposited at $U_B = -60$ V, shown by green triangles in Fig. 7.21a, the recalculated a_ψ^{hkl} according to Eq. (6.27) is illustrated by the dashed green line. Next, the Γ-independent part of the dependence of a_ψ^{hkl} on $\sin^2 \psi$ was determined, see Eq. (6.32) and Eq. (6.40), which yielded a linear function. Finally, the lattice parameters $a_\parallel$ and $a_\perp$ were determined from that linear function and a_0 was calculated from Eq. (6.22) assuming $\nu = 0.3$.

In that manner, the stress-free lattice parameters of the fcc-(Ti,Al)N phase from the three coating series were obtained and are shown in Fig. 7.21b. With increasing bias voltage the a_0 increased in all three coatings series. The comparison of a_0 for the coatings with different Ru content (series I to III) that were deposited at the same bias voltage showed no significant differences in a_0. The stress-free lattice parameters of the fcc-(Ti,Al)N phase was mainly

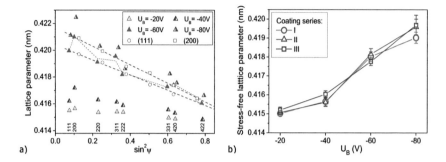

Fig. 7.21: The $\sin^2\Psi$ plot obtained for the Ti-Al-N / Al-Ti-Ru-N multilayers of series II deposited at various bias voltages and measured by GAXRD (triangles) (a) and the evaluation of the stress-free lattice parameter with increasing bias voltages of the coating series I to III (b). In (a) the open circles and boxes show the observed lattice parameters obtained from the interplanar spacing of the lattice planes (111) and (200) at different Ψ angles in the Ti-Al-N / Al-Ti-Ru-N multilayer coating deposited at $U_B = -60$ V. The green dashed line was recalculated according to Eq. (6.27).

affected by the bias voltage. Especially, the stress-free lattice parameters for the coatings deposited at high bias voltages ($U_B = -60$ V and $U_B = -80$ V) were higher than the value expected from the Vegard-like dependence (0.4165 nm) (see Eq. (7.1)) [82]. This indicates the segregation of Al from the fcc-(Ti,Al)N host structure. The Al which is not incorporated into fcc-(Ti,Al)N in the coatings deposited at high bias voltages forms wurtzitic phase (see page 95) as well as fcc-AlN or an Al-rich fcc-(Ti,Al)N phase that was deduced from the asymmetric peak broadening especially of the 200 diffraction line.

7.2.3 Multilayer architecture of the Ti-Al-N / Al-Ti-(Ru)-N multilayer coatings

TEM investigations of the cross section of the Ti-Al-N / Al-Ti-(Ru)-N multilayers revealed the multilayer architecture of the coatings. The TEM BF micrograph of the Ti-Al-N / Al-Ti-(Ru)-N multilayer coating of series III deposited at $U_B = -40$ V (Fig. 7.22a) illustrates exemplarily the coating architecture at the bottom of the coating. In order to improve the adhesion of the coating, a ~ 0.2 μm thick fcc-TiN adhesion layer was deposited onto the cemented carbide substrate. Next to the TiN adhesion layer a transition zone consisting of six repeating stacks composed of a ~ 12 nm thick Ti-Al-(Ru-)N and ~ 25 nm thick TiN layers were deposited. The actual Ti-Al-N / Al-Ti-(Ru)-N multilayers were grown after the transition zone. The fcc structure of the TiN adhesion layer and the transition zone was proven by SAED. Furthermore, SAED revealed the preferred orientation of the adhesion and transition layer of the coating. This is shown exemplarily in Fig. 7.22b and Fig. 7.22c for the coating of series III deposited at $U_B = -40$ V. SAED done in an area in the transition zone with a diameter of ~ 100 nm demonstrated, that the prevailing orientation is characterized by the $[212]_{fcc}$ direction of the fcc phase being parallel to the normal of the TEM foil and the $[10\bar{1}]_{fcc}$ direction inclined by $\sim 20°$ from the growth direction (see Fig. 7.22c). The weak reflections in the SAED in Fig. 7.22c that are highlighted by grey arrows could be attributed to a 2nd orientation of the fcc phase. This 2nd orientation is created by the rotation of the crystal by $35°$ around the $[10\bar{1}]$ direction so that the direction of the incident electron beam is parallel to the $[121]_{fcc}$ direction.

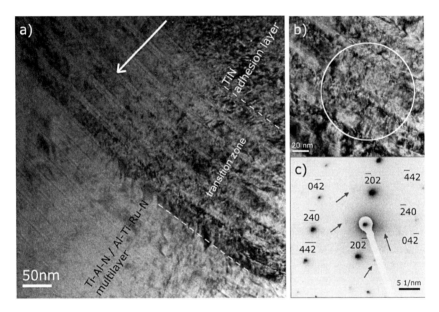

Fig. 7.22: TEM cross section of the Ti-Al-N / Al-Ti-Ru-N multilayer of series III deposited at $U_B = -40$ V showing the coating architecture at the bottom of the coating in a TEM BF image (a). The arrow in (a) denotes the growth direction. The white circle in (b) indicates the area of SAED in the transition zone. The SAED revealed the fcc-phase with $[212]_{fcc}$ direction being parallel with the normal of the TEM foil (c). The grey arrows in (c) highlight a 2nd orientation of the fcc-phase with the $[121]_{fcc}$ direction being parallel to the foil's normal.

In order to visualize the arrangement of the Ti-, Al- and Ru-rich regions within the coatings, STEM images in DF mode of the coatings in cross section were done. In case of a constant TEM foil thickness, the regions which are characterized by a high average atomic number (Z) appear brighter than the regions consisting of elements with lower Z in the STEM DF image.

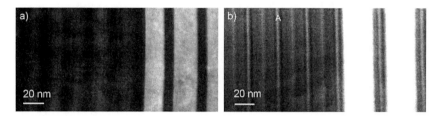

Fig. 7.23: STEM DF image of the cross section of the Ti-Al-N / Al-Ti-(Ru)-N multilayers of series I (a) and series III (b) that were deposited at U_B = -40 V. The growth direction is from the right to the left of the images.

In Fig. 7.23 STEM DF images of the Ru-free Ti-Al-N / Al-Ti-N multilayer of series I and the Ru containing Ti-Al-N / Al-Ti-Ru-N multilayer of series III, both deposited at U_B = -40 V, are shown. Both images illustrate the layered architecture of the coatings due to changes in the mean atomic weight within the coating. The bright layers with a thickness of ~25 nm at the right of both images correspond to the TiN layers of the transition zone. The length of the periodic motif of the actual Ti-Al-N / Al-Ti-(Ru)-N multilayers estimated from the STEM DF images is ~ 21 nm for the Ti-Al-N / Al-Ti-N multilayer of series I (Fig. 7.23a) and ~ 29 nm for the Ti-Al-N / Al-Ti-Ru-N multilayer of series III (Fig. 7.23b). The contrast changes in the STEM DF image of the Ru containing multilayer differ slightly from the Ru-free coating, because in Fig. 7.23b thin bright layers with a thickness of ~ 3 nm that are labeled as "A" are visible indicating a high Z region. This layer was also visible in the DF STEM image of the Ti-Al-N / Al-Ti-Ru-N multilayer of series III which was deposited at U_B = -80 V (Fig. 7.36b) and which is shown in Chapter 7.2.7.2. These kind of layers were not visible in Fig. 7.23a. Information about the changes in the chemical composition were obtained by EDS and EELS across the layer stacks along a line scan with a step size of 1 nm (see Fig. 7.24). The ratios of the metallic components were determined from the EDS analysis of the Ti Kα, Al Kα and Ru Kα lines and are shown in Fig. 7.24b and Fig. 7.24e for the Ti-Al-N / Al-Ti-(Ru)-N multilayers of series I and III, respectively. In addition to the determined Ru-ratio, the intensity of the Ru Kα line along the line scan is plotted in Fig. 7.24e as dotted blue line for the coating of series III. Since the Ru concentration is quite low ([Ru] / Σ[Me] $\leq$ 0.02), the EDS analysis routine implemented in *DigitalMicrograph^{TM}* [160] determined in some cases a Ru-ratio of [Ru] / Σ[Me] = 0 although a sufficient intensity of the Ru Kα line was measured e.g. at the distance of 80 nm for the coating of series III. However, the EDS analysis revealed, that the Ru enriched regions are located in the thin bright layers which were labelled as "A" in the Ru containing Ti-Al-N / Al-Ti-Ru-N multilayers. In this area also the highest Al concentration

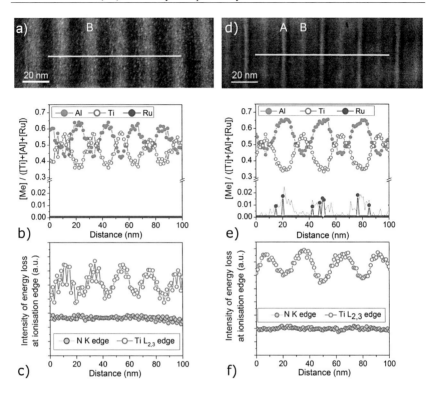

Fig. 7.24: DF STEM images of the Ti-Al-N / Al-Ti-(Ru)-N multilayers of coating series I (a) and III (d) deposited at U_B = -40 V. The white lines show the positions of the analytical TEM line scans. The results of the EDS (b,e) and EELS (c,f) analysis are given below the DF STEM images for the respective coating. The blue dotted line in (e) shows the intensity of the Ru Kα line which was observed using EDS analysis.

$[Al] / \Sigma[Me] \approx 0.65 \pm 0.01$ is found. This is in agreement with the target composition because the Al-rich target contained the Ru addition (see Chapter 7.2.1). No Ru was detected in the Ti-Al-N / Al-Ti-N multilayers of series I as shown in Fig. 7.24b. Due to that, the Al-rich layers are characterized by a low medium Z and appear dark in the DF STEM image (see Fig. 7.24a). The highest Ti concentration of $[Ti] / \Sigma[Me] \approx 0.57 \pm 0.01$ was found in the bright layers labelled as "B" in both coatings. The concentration profiles in Fig. 7.24b and Fig. 7.24e show that the Ti and Al concentration change gradually within the bi-layer period which is attributed to the 2-fold rotation causing the intermixing of the different target materials during deposition. The medium Al- and Ti-ratios averaged over three periodic layer stacks were (0.54 ± 0.01) and (0.46 ± 0.01), respectively in case of the Ti-Al-N / Al-Ti-N coating of series I deposited at U_B = -40 V and (0.55 ± 0.01) and (0.44 ± 0.01) in case of the Ti-Al-N / Al-Ti-Ru-N coating of series III deposited at U_B = -40 V. These results are in good agreement with the metal ratios determined by EPMA / WDS (see Table 7.1). The medium Ru content was underestimated by EDS in STEM, which is mainly due to limited sensitivity of the method for signal to noise ratio close to unity.

The presence of nitrogen was investigated by EELS. For this purpose, the electron energy loss (EEL) was measured in the range from 310 to 550 eV along the line scan with a step size of 1 nm. In that way, the EEL at the nitrogen K edge at ~ 400 eV (see Section 6.3.3.2) and at the titanium $L_{2,3}$ edge at 455.5 eV [168] could be recorded simultaneously. After background subtraction from the EEL spectra and removal of plural scattering effects, the intensity of the Ti core loss signal and the N core loss signal were determined. The results are shown in Fig. 7.24c and Fig. 7.24f for the Ti-Al-N / Al-Ti-(Ru)-N multilayer of series I and III deposited at U_B = -40 V, respectively.

EELS could verify the presence of nitrogen. The intensity of the core loss signal is relatively constant along the line scan, which is in accordance with Ref. [166], where a constant integral intensity of the N K edge ($I^N_{Ti_{1-x}Al_xN}$) for an Al concentration of $0.4 \leq x \leq 0.65$ was found (see also page 56).

Furthermore, the fluctuation of the Ti content across the layer stacks, that was observed by EDS analysis (see Fig. 7.24b and Fig. 7.24e), could be confirmed by the intensity of the Ti core loss signal of the EELS analysis.

As shown above TEM investigations revealed the multilayer architecture and the fluctuations of the metal concentration across the multilayer stacks. The change of the Al concentration across the multilayers might lead to a change of the crystal structure from fcc to wurtzite phase, since the Al ratio in the Al-rich layers reached 0.65 (see Fig. 7.24b,e) as observed by EDS. This Al ratio is close to the critical Al ratio of $x \approx 0.67$ that is reported in literature as the solubility limit of Al in fcc-TiN e.g. Refs. [43, 44]. Furthermore, this Al ratio lies within the Al concentration range where the formation of wurtzite phase was observed in CAE $Ti_{1-x}Al_xN$ monolayer coatings (cf. Fig. 7.3b) that were deposited similarly like the Ti-Al-N / Al-Ti-Ru-N multilayer coatings. In order to check, if the individual layers of the Ti-Al-N / Al-Ti-Ru-N multilayers are crystalline and if the crystal structure changes across the multilayer stacks, HRTEM and FFT was done. Fig. 7.25a shows an HRTEM image of the Ti-Al-N / Al-Ti-Ru-N multilayer coating of series III deposited at U_B = -40 V. The displayed area expands over one bi-layer period. The FFT of the whole HRTEM image revealed the fcc crystal structure of (Ti,Al)N (see Fig. 7.25b). Furthermore, continuous lattice fringes are visible in the entire HRTEM image. This indicates that the investigated coating is crystalline across the whole bi-layer period and that the fcc crystal structure as well as the orientation can be maintained over the Al-rich and Ti-rich layers. From the FFT it is also obvious that the crystallographic direction $\langle 113 \rangle_{fcc}$ is inclined by less than $10°$ from the growth direction of the film. This local orientation is in accordance with the global preferred orientation as observed by XRD pole figure measurements (see Chapter 7.2.6), because XRD revealed the $\langle 115 \rangle_{fcc}$ direction being parallel to the surface normal. The $\langle 115 \rangle$ direction is inclined by just $9°$ from the $\langle 113 \rangle$ direction in cubic structures.

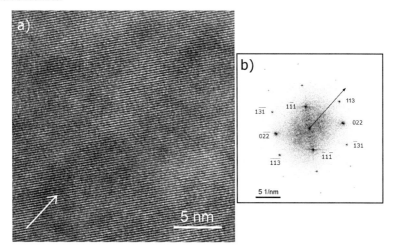

Fig. 7.25: HRTEM image (a) of the Ti-Al-N / Al-Ti-Ru-N multilayer coating of series III deposited at $U_B = -40$ V with the corresponding FFT (b) which shows the fcc phase with the $[21\overline{1}]_{fcc}$ direction being parallel to the direction of the primary electron beam. The arrows indicate the growth direction of the coating.

Nevertheless, the presence of traces of wurtzite phase in the as-deposited Ti-Al-N / Al-Ti-(Ru)-N multilayers were indicated by laboratory and synchrotron GAXRD experiments (see Fig. 7.19 and Fig. 7.20). Due to its small volume fraction in the coatings, its location is usually hard to identify by TEM methods. However, an interface between fcc-(Ti,Al)N and wurtzite phase could be identified by HRTEM in the Ti-Al-N / Al-Ti-N multilayer of series I, that was deposited at $U_B = -40$ V. The HRTEM image in Fig. 7.26a shows a column boundary which separates both phases. The FFT (Fig. 7.26b) that was done in the upper left part of the image and which is marked by the yellow square, revealed the fcc-(Ti,Al)N phase. The diffraction spots in the FFT (Fig. 7.26b) correspond to the black labeled reflection given in the simulated electron diffraction pattern in Fig. 7.26f. According to that, the $(001)_{fcc}$ plane of the fcc phase is parallel to the TEM foil and the $[\overline{1}\overline{1}0]_{fcc}$ direction is inclined by ~12 ° from the

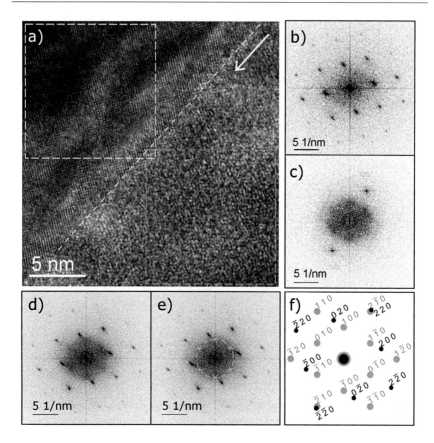

Fig. 7.26: HRTEM image (a) of the Ti-Al-N / Al-Ti-N multilayer coating of series I deposited at $U_B = -40$ V. The white arrow indicates the growth direction of the coating and the white dashed line illustrates the column boundary. The FFT in (b) and (c) correspond to the yellow and orange marked squares. The FFTs in (d) and (e) are from the whole HRTEM image showing that the $(001)_{fcc}$ plane of the fcc phase is parallel to the TEM foil plane. Additionally, diffuse diffraction spots are visible which can be attributed to the wurtzite phase that is oriented with the $(001)_w$ plane parallel to the TEM foil plane. The diffuse diffraction spots from the wurtzite phase are highlighted in (e). The simulated electron diffraction pattern is given in (f) where the black and orange marked diffraction spots correspond to the fcc phase and wurtzite phase, respectively. In the simulation the $[001]_{fcc}$ direction of the fcc phase and the $[001]_w$ direction of the wurtzite phase are parallel to the direction of the primary electron beam.

growth direction. The FFT of the whole HRTEM image is given in Fig. 7.26d. Here the reflection coming from the fcc phase, which is orientated with the $(001)_{fcc}$ plane parallel to TEM foil plane, are clearly visible (cf. simulation in Fig. 7.26f). Additional diffuse and broad reflections are visible in the FFT of Fig. 7.26d. These reflections appear at a constant distance from the origin (000) and with 6-fold symmetry. The FFT in Fig. 7.26c obtained from the lower right part of the HRTEM image that is marked by the orange square (see Fig. 7.26c) reveals that these diffuse reflections originate from the right side of the column boundary. Their distance and positions are highlighted in Fig. 7.26e by an orange circle and cross, respectively. The reciprocal lattice distances of these reflections that were estimated from the FFT are ~ 3.7 nm^{-1}. Considering this distance as well as the 6-fold symmetry, the fcc-(Ti,Al)N can be excluded as possible origin of these reflections. These reflections can be rather explained by w-AlN that is oriented with the $(001)_w$ plane parallel to the TEM foil plane as illustrated by the orange labeled reflections shown in the simulation in Fig. 7.26f. Due to the diffuse appearance of the w-AlN diffraction spots in (Fig. 7.26c), it is assumed that the wurtzite phase is characterized by small distorted crystallites in the investigated sample area displayed in the HRTEM image in Fig. 7.26. Furthermore, the wurtzite phase is oriented with the $(\overline{2}10)_w$ plane parallel to the $(\overline{1}10)_{fcc}$ plane of the fcc phase which corresponds to $[\overline{1}00]_w \parallel [\overline{1}10]_{fcc}$. Exactly this relationship between fcc and wurtzite phase was observed previously in the Ti$_{0.44}$Al$_{0.56}$N monolayer coating deposited at $U_B = -40$ V as presented in Chapter 7.1.5 and Fig. 7.7. The schematic presentation of such an interface is given in Fig. 7.8 and Fig. 7.9.

7.2.4   Residual stress in the as-deposited Ti-Al-N / Al-Ti-(Ru)-N multilayer coatings

The residual stress and the macroscopic lattice strain of the fcc-(Ti,Al)N-phase in the Ti-Al-N / Al-Ti-(Ru)-N multilayers were determined from the dependence of a_ψ^{hkl} on sin$^2 \psi$ as shown in Fig. 7.21a. Due the presence of the anisotropy of the elastic lattice deformation, the Γ-independent part was calculated (see Eq. (6.32) and Eq. (6.40)) which yielded a linear function of a_ψ^{hkl} vs. sin$^2 \psi$. From that linear function the macroscopic lattice strain was calculated according to Eq. (6.23) and the residual stress was determined using Eq. (6.24) assuming a Young's modulus of $E = 500$ GPa. The results for all three Ti-Al-N / Al-Ti-Ru-N multilayer series are shown in Fig. 7.27.

All coatings are characterized by a compressive stress state. The lowest macroscopic lattice strain was observed for the coatings deposited at the lowest bias voltage of -20 V. The lattice strain corresponds to a compressive stress state of ~ 1 GPa. With increasing bias voltage from $U_B = -20$ V to $U_B = -60$ V the macroscopic lattice strain increased nearly linearly. The compressive stress reached ~ 6 GPa at $U_B = -60$ V. When the bias voltage was increased further to $U_B = -80$ V, the macroscopic lattice strain and the compressive stress increased only slightly reaching approx. ~ 7 GPa. The development of the macroscopic lattice strain of the fcc-(Ti,Al)N phase as function of the bias voltages was similar for the Ti-Al-N / Al-Ti-(Ru)-N multilayer coatings with different Ru indicating that the Ru addition has no significant influence on the stress evolution in the as-deposited state.

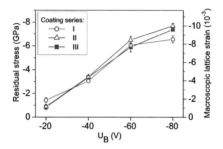

Fig. 7.27: Influence of the bias voltage on the residual stress and macroscopic lattice strain of the fcc-(Ti,Al)N phase in the Ti-Al-N / Al-Ti-Ru-N multilayers with different Ru content. The lines are guide for the eyes.

The increase of the compressive stress with rising negative bias voltage can be attributed to intensified defect generation e.g. displaced atoms due to direct and recoil implantation of film atoms [198] caused by a higher kinetic energy of the bombarding ions due to the increased potential difference (ΔV_{sheath}) between the plasma and the substrate. The marginal increase of the compressive stress when the bias voltage was increased from $U_B = -60$ V to $U_B = -80$ V indicated a partial annihilation of lattice defects, e.g. relaxation of displaced atoms, due to local heating of the deposited coating as a result of bombarding ions which leads just to a minor stress increase.

7.2.5   Hardness of the as-deposited Ti-Al-N / Al-Ti-(Ru)-N multilayer coatings

The indentation hardness of the Ti-Al-N / Al-Ti-(Ru)-N multilayers is shown in Fig. 7.28. The hardness evolution vs. the bias voltage can be classified into two groups. The first group comprises the coatings deposited at low bias voltages (-20 V and -40 V). Their hardness lies in the range between 30 and 32 GPa. The second group contains the coatings deposited at high bias voltage (-60 V and -80 V). This group is characterized by the highest hardness values lying between 33 and 35 GPa. The increasing hardness with rising bias voltage indicates the contribution of the higher compressive stress to the hardness enhancement in the coatings. It

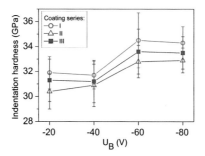

Fig. 7.28: Influence of the bias voltage on the indentation hardness of Ti-Al-N / Al-Ti-(Ru) with different Ru content.

seems that the Ru addition had no remarkable effect on the hardness. The indentation hardness differs for different Ru additions within the range of errors.

7.2.6   Preferred orientation of the fcc-(Ti,Al)N crystallites

The $\{111\}_{fcc}$, $\{100\}_{fcc}$ and $\{110\}_{fcc}$ pole figures were measured for all Ti-Al-N / Al-Ti-(Ru)-N multilayer coatings in order to determine the preferred orientation of the fcc-(Ti,Al)N crystallites. Examplarily the pole figures which were measured for the Ti-Al-N / Al-Ti-(Ru)-N multilayer coating of series I deposited at U_B = -60 V are shown in Fig. 7.29. Similarly to the pole figures shown in Fig. 7.29, the pole figures measured in the multilayer coatings of the three coating series showed rings with a cylindrical intensity profile at certain polar angles. This indicated an out-of-plane preferred orientation. The pole figures of the fcc-(Ti,Al)N crystallites revealed that neither the majority of the $\{111\}_{fcc}$ nor the $\{100\}_{fcc}$ nor the $\{110\}_{fcc}$ planes are parallel to the coating's surface. The majority of these planes were inclined from the normal direction. The lowest inclination was found for $\{100\}_{fcc}$ planes. In order to determine the inclination angle between the $\langle 100 \rangle_{fcc}$ direction and the surface normal direction, the intensity profile across the cross section of the $\{100\}_{fcc}$ pole figure was fitted by two Gaussian functions (see Eq. (6.46)) as shown in Fig. 7.29d. The inclination angle of the $\langle 100 \rangle_{fcc}$ direction from the surface normal direction was determined from the center of the Gauss fit and the degree of the out-of-plane preferred orientation was deduced from the FWHM of the Gauss fit.

In case of the example shown in Fig. 7.29, the $\langle 100 \rangle_{fcc}$ direction is inclined by ~ 17° from the surface normal. The inclination angle of the $\langle 100 \rangle_{fcc}$ direction from the surface normal direction for all three coating series is given in Fig. 7.30a. The coatings deposited at bias voltages

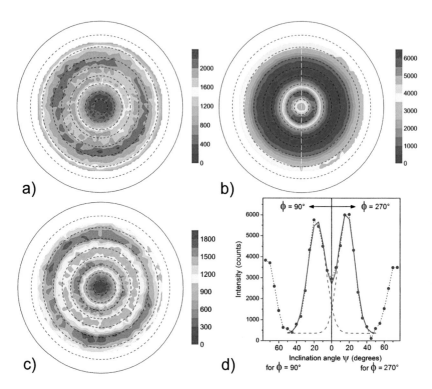

Fig. 7.29: Pole figures {111} (a), {100} (b) and {110} (c) of the fcc-(Ti,Al)N phase in the Ti-Al-N / Al-Ti-N multilayer coating of series I deposited at $U_B = -60$ V. The dashed grid lines correspond to the inclination angle ψ with an interval of 10°. The white dashed line in (b) notes the position of the intensity profile that was fitted by two Gaussian functions. This intensity profile is given in (d) where the blue symbols denote the measured intensity and the red dashed curves show the two Gaussian functions.

of -20 V, -40 V and -60 V are characterized by an inclination of the $\langle 100 \rangle_{fcc}$ direction in the range of 14 to 17°. At the highest bias voltage of -80 V the inclination of the $\langle 100 \rangle_{fcc}$ direction from the surface normal direction increased noticeably and ranged between 20 and 24°. This trend of the inclined $\langle 100 \rangle_{fcc}$ direction from the surface normal direction was similar for the three coating series with different Ru content.

The degree of the out-of plane preferred orientation was similar for all coatings (see Fig. 7.30b), neither the bias voltage nor the Ru content in the film seemed to have an influence.

The inclination of the $\langle 100 \rangle_{fcc}$ direction by 14° to 24° from the surface normal brings high indices lattice planes parallel to the sample surface. For the low inclination of ~14°, the {511}$_{fcc}$

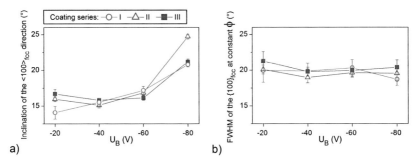

a) b)

Fig. 7.30: Analysis of the {100}$_{fcc}$ pole figure showing the inclination of the $\langle 100 \rangle_{fcc}$ direction from the sample surface (a) and the FWHM of the {100}$_{fcc}$ pole determined at a constant angle ϕ (b) as function of the bias voltage U$_B$.

lattice planes are nearly parallel to sample surface. The deviation of the $\langle 511 \rangle_{fcc}$ direction from the surface normal direction is just $\sim 2°$. In the case of a perfect {511}$_{fcc}$ out-of-plane texture, the {100}$_{fcc}$ planes are inclined by 16°, 79°, the {110}$_{fcc}$ planes are inclined by 35°, 57°, 74° and the {111}$_{fcc}$ planes are inclined by 39°, 56° and 70° from the sample surface as calculated by Eq. (6.47) (see Table 7.2). The theoretical pole figures simulated for a perfect {511}$_{fcc}$ out-of-plane texture visualizing the intensity rings at the respective polar angles are shown in the Appendix in Fig. A. 1. Due to the slight deviation of the measured texture from the {511}$_{fcc}$ out-of plane texture as well as due to the degree of the out-of plane texture shown in Fig. 7.30b and the step size of the inclination angle ψ of 5° that was used during the measurement of the pole figures, the expected individual intensity rings in the {111}$_{fcc}$ pole figure at the polar angles 39°, 56°, 70° appear as a broad intensity ring in the range of $\sim 35 - 70°$ in the measured {111}$_{fcc}$ pole figure. The measured intensity rings in the {110}$_{fcc}$ pole figure at the polar angles $\sim 35°$ and $\sim 60°$ can be attributed to the expected individual intensity rings at 39°, 56° for an ideal {511}$_{fcc}$ texture.

Table 7.2: Angles between the {100}$_{fcc}$, {110}$_{fcc}$ and {111}$_{fcc}$ lattice planes and the sample surface in case of (511)$_{fcc}$ and (311)$_{fcc}$ texture as calculated by Eq. (6.47).

| | | (001) | (101) | (1$\bar{1}$0) | | (0$\bar{1}$1) | | (1$\bar{1}$1) | |
	(100)	(010)	(110)	(10$\bar{1}$)	(011)	(01$\bar{1}$)	(111)	(11$\bar{1}$)	(1$\bar{1}\bar{1}$)
(511)	15.79°	78.90°	35.26°	57.02°	74.21°	90.00°	38.94°	56.25°	70.53°
(311)	25.23	72.45	31.48	64.76°	64.76°	90.00°	29.50°	58.52°	79.98°

The pole figures of the Ti-Al-N / Al-Ti-(Ru)-N multilayer coatings of all three series deposited at U_B = -20 V, -40 V and -60 V were similar to the ones shown in Fig. 7.29.

The pole figures of the Ti-Al-N / Al-Ti-(Ru)-N multilayer coatings of all three series deposited at U_B = -80 V showed also intensity rings at certain polar angles. This is shown exemplarily for the Ti-Al-N / Al-Ti-N multilayer coating of series I deposited at U_B = -80 V in Fig. 7.31. In the coatings deposited at U_B = -80 V the $\langle 100 \rangle_{fcc}$ direction was inclined by ~ 21 to $\sim 24°$ from the surface normal (see Fig. 7.30a).

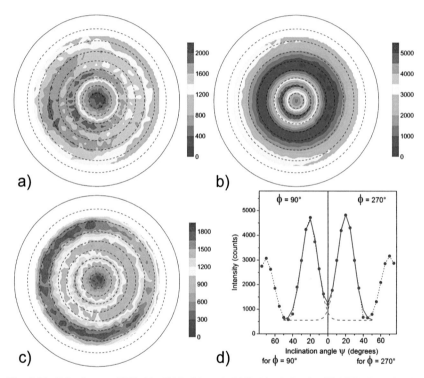

a)

b)

c)

d)

Fig. 7.31: Pole figures {111} (a), {100} (b) and {110} (c) of the fcc-(Ti,Al)N phase in the Ti-Al-N / Al-Ti-N multilayer coating of series I deposited at U_B = -80 V and the intensity profile of the {100} pole figure that was fitted by two Gaussian functions (d). The symbols in (d) have the same meaning like in Fig. 7.29. The dashed grid lines correspond to the inclination angle ψ with an interval of 10°.

Furthermore, intensity rings in the $\{111\}_{fcc}$ pole figure at the polar angles $\sim 35°$ and ~ 50 - $60°$ and intensity rings at the polar angles $\sim 30°$ and $\sim 65°$ in the measured $\{110\}_{fcc}$ pole figure were observed. Thus, it was concluded that the $\{311\}_{fcc}$ lattice planes tend to align nearly parallel to the sample surface at the highest bias voltage. Because in the case of a perfect $\{311\}_{fcc}$ out-of-plane texture (see Appendix Fig. A. 1b), the $\{100\}_{fcc}$ planes are inclined by 25°, 72°, the $\{110\}_{fcc}$ planes are inclined by 31°, 65°, 90° and the $\{111\}_{fcc}$ planes are inclined by 29°, 59° and 80° from the sample surface as calculated by Eq. (6.47) (see Table 7.2).

Several theories about the texture formation in thin films are proposed in literature (see e.g. Ref. [216]). Among these theories, the influence of the ion channelling direction [216] and the competition between strain E_{strain}^{hkl} and surface $E_{surface}^{hkl}$ energies [217] are the most frequently reported ones.

According to the theory of the ion channelling direction, the lattice planes with easy channelling directions and lower sputtering yield have the highest probability to survive. This effect was especially observed when the growing film is exposed to high ion energy at low-ion-flux (ion to metal ratio ≤ 20 %) [218, 219]. Because in case of wider channelling directions, the energy of the impinging ions is distributed over larger volumes and the lattice planes will be less distorted. In fcc-TiN the relative numbers of atom columns per unit area of a^2 for the $[100]_{fcc}$, $[110]_{fcc}$ and $[111]_{fcc}$ direction correspond to $1 : \sqrt{2} : \sqrt{3}$ (see Appendix Fig. A. 2). Thus, the $[100]_{fcc}$ direction is the most open channelling direction whereas the densest array of atoms is found in the $[111]_{fcc}$ direction. If the ions' incident direction is parallel to the normal of the surface, the $\{100\}_{fcc}$ preferred orientation is formed at high ion energy and low-ion-flux as observed in Refs. [216, 219]. The correlation between the open channelling directions and the texture formation were verified in TiN films grown by magnetron sputtering [218, 219] and ion assisted deposition [216]. Since during CAE deposition nearly all species arriving at the substrate are ionized and highly energetic, the formation of a preferred orientation with open channelling directions parallel to the direction of the incoming ion flux could be expected. However, due to the used substrate rotation for the deposition of the Ti-Al-N / Al-Ti-Ru-N multilayers, the direction of the bombarding ions is changing continuously. Thus, the role of the open channeling direction may constitute an underpart.

Considering the theory about the competition between strain E_{strain}^{hkl} and surface $E_{surface}^{hkl}$ energies the $\{111\}_{fcc}$ texture would be expected in thick TiN coatings, as it was observed e.g. in Ref. [220] for a fcc-TiN coating deposited using CAE. The formation of the $\{111\}_{fcc}$ texture was also observed in a 4 μm thick fcc-TiN coating that was deposited by CAE at $U_B = -80$ V with similar deposition parameters in the same deposition apparatus like the Ti-Al-N / Al-Ti-(Ru)-N multilayer coatings (see Appendix Fig. A. 3).

The formation of a $\{111\}_{fcc}$ texture in thick TiN coatings is attributed to the fact that both energy contributions depend on the crystallographic direction. Fcc-TiN is characterized by an anisotropic elastic lattice deformation with the Young's modulus obeying $E_{111} < E_{110} < E_{100}$.

The strain energy contribution is determined by the mean elastic moduli acting in the plane (see Eq. (A.2)) [217]. Considering this, the strain energy will be greater for the {100} than for the {111} and {110} orientation because the directions $\langle 110 \rangle_{fcc}$ and $\langle 111 \rangle_{fcc}$ with medium and low Young's moduli are lying in the {111}$_{fcc}$ and {110}$_{fcc}$ planes, whereas the directions $\langle 100 \rangle_{fcc}$ and $\langle 110 \rangle_{fcc}$ with high and medium Young's moduli are lying in the {100}$_{fcc}$ plane.

The surface energy for the {100}, {110} and {111} planes obey the sequence $E^{100}_{surface} < E^{110}_{surface} < E^{111}_{surface}$ according to Refs. [217, 221, 222]. Tholander *et al.* [223] showed recently by DFT calculations that the surface energy obeys the sequence $E^{100}_{surface} < E^{111}_{surface} < E^{110}_{surface}$ if a N - terminated {111} surface is considered. However, all authors found that the {100} surface has the lowest surface energy [217, 221, 222, 224]. For low coating thicknesses, the surface energy is the dominating energy contribution and hence the {100}$_{fcc}$ texture will form. At a high coating thickness e.g. > 300 nm [217] the strain energy contribution will be the dominating one and thus the {111} texture will be favoured.

The thickness of the Ti-Al-N / Al-Ti-(Ru)-N multilayers is in the range of 4 to 5 μm and the fcc-(Ti,Al)N phase in the multilayers is strained (cf. Chapter 7.2.4) which suggests that the contribution of the strain energy is dominating in the multilayers.

However, the anisotropy of the elastic lattice deformation in coatings based on Ti$_{1-x}$Al$_x$N depends on the Al content. Tasnadi *et al.* [215] showed by *ab initio* calculations that the ratio E_{111}/E_{100} increases from 1 for x = 0.28 to approx. 2 for x = 1, which means $E_{100} < E_{111}$ or anisotropy factor $A > 1$ (see Eq. (6.41)) for x > 0.28. This implies that the strain energy for Ti$_{1-x}$Al$_x$N coatings with x > 0.28 is lower for the {100} planes than for the {111} planes.

The presence of the anisotropy of the elastic lattice deformation with $A > 1$ was found in the Ti-Al-N / Al-Ti-(Ru)-N multilayer coatings (see Fig. 7.21 page 98) and the average [Al] / Σ[Me] ratio is approx. 0.53. This means that E_{100} is smaller than E_{111} in the Ti-Al-N / Al-Ti-(Ru)-N multilayer coatings. Thus, it would be expected that a {100}$_{fcc}$ texture forms in these coatings.

However, the {100}$_{fcc}$ planes in the Ti-Al-N / Al-Ti-(Ru)-N multilayers are inclined from the sample surface by 14° to 24° bringing the low index lattice planes {311}$_{fcc}$ and {511}$_{fcc}$ nearly parallel to the sample surface. A similar preferred orientation was observed by Karimi *et al.* [225] in Ti-Al-N coatings that were deposited by CAE using Ti$_{50}$Al$_{50}$ targets. The origin of such texture formation is not understood yet.

7.2.7   Thermal stability of Ti-Al-N / Al-Ti-(Ru)-N multilayers

7.2.7.1 Thermally activated microstructure changes in the Ti-Al-N / Al-Ti-(Ru)-N multilayer coatings

In order to study the influence of the Ru addition and the bias voltage on the thermal stability of the Ti-Al-N / Al-Ti-(Ru)-N multilayers, each coating was annealed at 450 °C, 650 °C, 850 °C and 950 °C for 60 min in argon atmosphere. *Ex situ* GAXRD was done after each annealing step and the GAXRD patterns were analysed with respect to the stress-free lattice parameters and the macroscopic lattice strains of the fcc phases. The indentation hardness was measured after each annealing step, in order to correlate the microstructure changes, induced by the thermal treatment, with the coating properties. The obtained results are shown in Fig. 7.32 and in Fig. A. 5 (see Appendix) for the Ti-Al-N / Al-Ti-(Ru)-N multilayers with three different Ru additions that were deposited at $U_B = -40$ V and $U_B = -80$ V and at $U_B = -20$ V and $U_B = -60$ V, respectively. Since the coatings deposited at $U_B = -40$ V and $U_B = -80$ V represent the general changes taking place in the coatings deposited at low and at high bias voltages the following discussion is focused on these both bias voltages. The microstructure changes taking place in the coatings deposited at $U_B = -20$ V and $U_B = -60$ V are shown in the Appendix (see Fig. A. 5).

The comparison of the evolution of the stress-free lattice parameter, the macroscopic lattice strain and the hardness revealed that the coatings which were deposited at the same bias voltage but contained different Ru additions did not differ significantly in the evolution of the microstructure. This indicates that during the annealing in argon atmosphere the Ru addition had no remarkable influence on the thermal stability of the Ti-Al-N / Al-Ti-(Ru)-N multilayers. Nevertheless, a beneficial effect of the Ru concentration on the thermal stability or hardness is possible during the application of the Ti-Al-N / Al-Ti-(Ru)-N multilayers in cutting processes as reported in Ref. [101]. During cutting, the coatings are exposed to high temperatures for less than 30 min [101] and the decomposition of the metastable fcc-(Ti,Al)N phase takes place simultaneously with the oxidation of the samples since cutting is usually done in air. Additionally, the lifetime of the coated inserts in cutting tests is influenced by tribological and tribochemical processes taking place due to the contact between the coating and the work piece. The impact of these processes is eliminated within the annealing experiments performed within this work. However, during annealing in argon atmosphere the influence of the bias voltage on the thermal stability of the Ti-Al-N / Al-Ti-(Ru)-N multilayers was evident. Thus, in the following, the discussion on the thermal stability of Ti-Al-N / Al-Ti-(Ru)-N multilayers is focused on the bias voltage and on the multilayer with the highest Ru addition (series III).

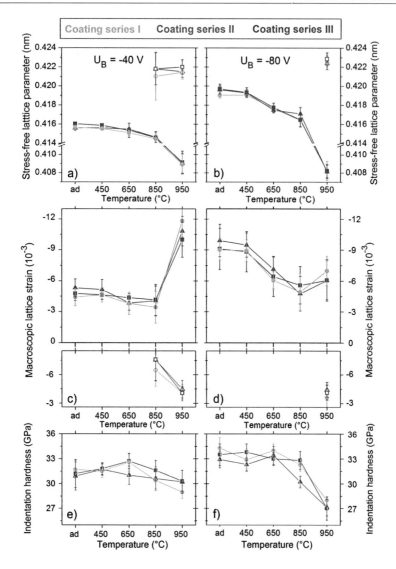

Fig. 7.32: Microstructure evolution of the Ti-Al-N / Al-Ti-(Ru)-N multilayers of the three coating series deposited at $U_B = -40$ V (left column) and $U_B = -80$ V (right column) in the course of the thermal treatment showing the stress-free lattice parameter (a,d) and macroscopic lattice strain (b,e) of the fcc-phases as well as the indentation hardness (c,f). The filled and open symbols represent the microstructure parameters of the fcc-(Ti,Al)N and nearly Al-free fcc-(Ti,Al)N, respectively. After 950°C the filled symbols represent rather the nearly Ti-free fcc-(Al,Ti)N phase.

Regarding Fig. 7.32 and Fig. A. 5, no apparent changes in the stress-free lattice parameter and macroscopic lattice strain of the fcc-(Ti,Al)N phase became apparent after annealing at **450°C.** No significant changes in the microstructure were expected since the deposition using CAE was done at the same temperature.

The subsequent annealing step at **650°C** resulted in changes of the microstructure. The stress-free lattice parameter and macroscopic lattice strain of the fcc-(Ti,Al)N phase decreased. These effects are especially pronounced in the coatings deposited at high bias voltages ($U_B = -60$ V and $U_B = -80$ V), see Fig. A. 5b,d and Fig. 7.32b,d.

This can be attributed to defect annihilation and the incorporation of Al into fcc-(Ti,Al)N. For example in the multilayer coating of series III deposited at $U_B = -80$ V the stress-free lattice parameter decreased to 0.4178 nm after annealing at 650 °C (see Fig. 7.32b) which indicates an Al content of x = 0.45 according to the Vegard-like dependence (Eq. (7.1)). However, after annealing at 650°C, the stress-free lattice parameter and the macroscopic lattice strain are still higher in the coatings deposited at $U_B = -60$ V and $U_B = -80$ V than in the coatings deposited at $U_B = -20$ V and $U_B = -40$ V (cf. Fig. A. 5 and Fig. 7.32).

In all coatings the onset of the decomposition was observed after annealing at **850°C** as shown exemplarily in Fig. 7.33 for the coatings deposited at $U_B = -40$ V and $U_B = -80$ V. The Ti-rich and Al-rich fcc-(Ti,Al)N phases formed and the fraction of the wurtzite phase increased slightly. These three phases existed beside the major fcc-(Ti,Al)N phase. This is visible in the diffraction patterns obtained in a laboratory GAXRD experiment using Cu radiation (see Fig. 7.33) as well as in the *ex situ* GAXRD patterns obtained by synchrotron radiation with a wavelength of $\lambda = 0.10781$ nm (see Fig. 7.34). Additionally, the fitted diffraction patterns of the Ti-Al-N / Al-Ti-Ru-N multilayer coatings of series III deposited at $U_B = -40$ V and $U_B = -80$ V showing the present phases after annealing at 850°C and 950°C are given in the Appendix (Fig. A. 4). In case of the coatings deposited at $U_B = -40$ V the stress-free lattice parameter of the Ti-rich fcc-(Ti,Al)N phase (see open symbols in Fig. 7.32a) indicated an aluminium incorporation of x ~ 0.16. The Ti-rich fcc-(Ti,Al)N phase as well as the fcc-(Ti,Al)N major phase were under compressive stress (Fig. 7.32c). The stress-free lattice parameter of the major fcc-(Ti,Al)N phase decreased to 0.4146 nm indicating the Ti depletion. After annealing at 850°C the stress-free lattice parameter of the Ti-rich fcc-(Ti,Al)N phase present in the coatings deposited at $U_B = -80$ V could not be determined since just the 111 and 200 diffraction lines of this phase (Ti-rich fcc-(Ti,Al)N) could be fitted (see Fig. A. 4). The further reduction of a_0 of the major fcc-(Ti,Al)N phase to 0.4165 nm implied the Ti depletion. The major fcc-(Ti,Al)N phase was still under compressive stress as indicated by the macroscopic lattice strain (Fig. 7.32d).

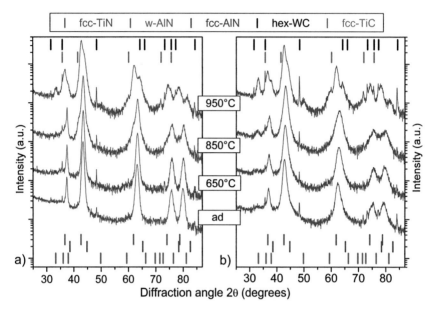

Fig. 7.33: Parts of the *ex situ* GAXRD patterns of the Ti-Al-N / Al-Ti-Ru-N multilayers of series III that were deposited at $U_B = -40$ V (a) and $U_B = -80$ V (b) and annealed in argon atmosphere for 60 min at the respective temperature.

After annealing at **950°C**, the nearly complete decomposition of the fcc-(Ti,Al)N phase into the wurtzite phase, Ti-rich fcc-(Ti,Al)N and almost Ti-free fcc-AlN was observed in all coatings. This is exemplarily shown by the GAXRD patterns for the coatings of series III deposited at $U_B = -40$ V and $U_B = -80$ V in Fig. 7.33 and in the fitted profiles in Fig. A. 4. The analysis of the stress-free lattice parameters indicated nearly Al-free fcc-TiN and nearly Ti-free fcc-AlN (see Fig. 7.32, Fig. A. 5). Both phases were under compressive stress.

The microstructure analysis revealed that the coatings deposited at high U_B decomposed to a further extent than the coatings that were deposited at low U_B. This was concluded from (i) a_0 of the Ti-rich fcc-(Ti,Al)N phase, (ii) a_0 of the Al-rich fcc-(Al,Ti)N phase, (iii) the intensity of the diffraction lines of the wurtzite phase in the GAXRD patterns and (iv) the hardness reduction. The individual items can be explained with the aid of Fig. 7.32 and Fig. 7.33 as follows.

(i)   The higher stress-free lattice parameter of the Ti-rich fcc-(Ti,Al)N phase of the coatings deposited at $U_B = -80$ V as compared to $U_B = -40$ V (see open symbols in Fig. 7.32a,b) signifies a lower remaining Al incorporation in fcc-(Ti,Al)N at high U_B than at low U_B after annealing at 950°C. According to Eq. (7.1), the Al concentration in the Ti-rich fcc-$Ti_{1-x}Al_xN$ phase of the coating series III is $x = 0.15$ for $U_B = -40$ V and $x = 0.09$ for $U_B = -80$ V.

(ii)   The lower stress-free lattice parameter of the Al-rich fcc-(Al,Ti)N phase of the coatings deposited at $U_B = -80$ V as compared to $U_B = -40$ V (see filled symbols in Fig. 7.32a,b) indicates a lower Ti concentration in the Al-rich fcc-(Al,Ti)N phase as compared to the coatings that were deposited at $U_B = -40$ V.

Both observations (i, ii) denote that the separation of aluminium nitride and titanium nitride progressed to a further extent in the coatings that were deposited at high U_B.

(iii)   After annealing at 950°C the diffraction lines of the wurtzite phase observed in the coatings deposited at $U_B = -80$ V possess a higher intensity than the ones observed in the coatings deposited at $U_B = -40$ V as shown exemplarily for the coating series III in Fig. 7.33 or in Fig. A. 4.

(iv)   The indentation hardness of the multilayer coatings deposited at $U_B = -40$ V decreased only slightly below the hardness level of the as-deposited state from approx. 32 GPa to approx. 30 GPa after annealing at 950°C. In contrast to that, the hardness of the multilayer coatings deposited at $U_B = -80$ V dropped from approx. 33 GPa to approx. 27 GPa (Fig. 7.32e,f). This

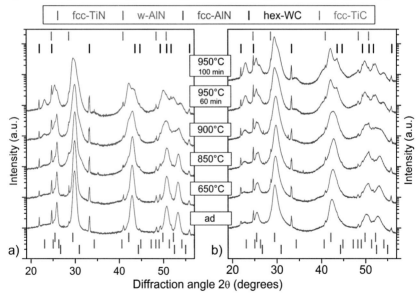

Fig. 7.34: Parts of the *ex situ* GAXRD patterns of the Ti-Al-N / Al-Ti-Ru-N multilayers of series III that were deposited at $U_B = -40$ V (a) and $U_B = -80$ V (b) and annealed in vacuum. The patterns were obtained with synchrotron radiation with a wavelength of 0.10781 nm after cooling from the annealing temperature to 100°C.

obvious hardness reduction at $U_B = -80$ V can be attributed to a higher volume fraction of the wurtzite phase than in the coatings deposited at $U_B = -40$ V which was also implied by the higher intensity of the diffraction lines of the wurtzite phase (see iii).

The accelerated decomposition of the Ti-Al-N / Al-Ti-Ru-N multilayer coatings that were deposited at high U_B was also found for the $Ti_{1-x}Al_xN$ coatings (see Section 7.1.7.2) and confirms the effect of the bias voltage on the thermal stability of Ti-Al-N based coatings.

A similar dependence of the microstructure evolution as shown in Fig. 7.32 was found for the *ex situ* GAXRD experiments done with synchrotron radiation after annealing in vacuum (see Fig. 7.35). Due to the experimental set up (see Chapter 6.2.2), the GAXRD patterns could be recorded just up to 83° 2θ. Thus, the lattice parameter a_ψ^{hkl} could be acquired just in the $\sin^2 \psi$-range between 0.03 and 0.39. As a consequence of this, the error for the calculation of the stress-free lattice parameter and the macroscopic lattice strain increases especially in coatings with a high residual stress level. This explains the rather large size of the error bars in Fig. 7.35.

However, the influence of the bias voltage on the thermal stability of the Ti-Al-N / Al-Ti-Ru-N multilayer coatings (here just studied on series III) was also evident from the *ex situ* GAXRD experiments done with synchrotron radiation in vacuum. The coating of series III that was deposited at $U_B = -80$V was characterized by an accelerated decomposition as compared to the coating deposited at $U_B = -40$ V, because the Ti-rich fcc-(Ti,Al)N phase in the coating $U_B = -80$ V contained a lower Al concentration ($x \approx 0.06$) than the coating $U_B = -40$ V ($x \approx 0.23$). Additionally, the Al-rich fcc-(Al,Ti)N phase in the coating $U_B = -40$ V indicated a higher Ti incorporation than in the coating $U_B = -80$ V (see Fig. 7.35a,b). Furthermore, a lower volume fraction of the wurtzite phase is expected in the coating $U_B = -40$ V than in the coating $U_B = -80$ V, since the diffraction lines of the wurtzite phase are less intensive (see Fig. 7.34) and the hardness reduced just from (31 ± 1.7) GPa to (29 ± 0.8) GPa in the coating $U_B = -40$ V. After the complete thermal treatment in vacuum, the Ti-Al-N / Al-Ti-Ru-N multilayer coating of series III that was deposited at $U_B = -80$ V was subjected to TEM analysis and is shown in the next Section 7.2.7.2.

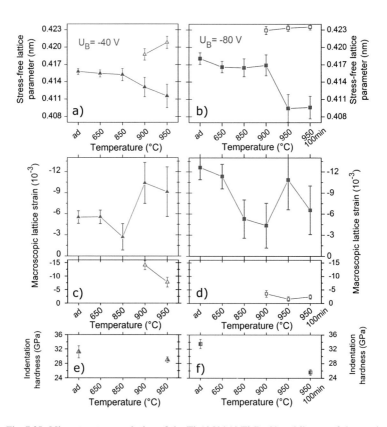

Fig. 7.35: Microstructure evolution of the Ti-Al-N / Al-Ti-Ru-N multilayers of the coating series III deposited at $U_B = -40$ V (left column) and $U_B = -80$ V (right column) in the course of the annealing in vacuum showing the stress-free lattice parameter a_0 (a,b) and macroscopic lattice strain ε (c,d) of the fcc-phases as well as the indentation hardness (e,f). The microstructure parameters a_0 and ε were obtained from the synchrotron experiments ($\lambda = 0.10781$ nm) after cooling from the annealing temperature to 100°C. The hardness of the coatings were determined prior and after the thermal cycle.

7.2.7.2   Thermally activated microstructure changes in the Ru-rich Ti-Al-N / Al-Ti-Ru-N multilayer coating deposited at $U_B = -80$ V as seen by analytical TEM

A TEM cross section sample was prepared from the Ti-Al-N / Al-Ti-Ru-N multilayer coating of series III which was deposited at $U_B = -80$ V and subjected to a thermal cycle up to 950°C (see Fig. 6.2b) and a total heating time of 100 min at 950°C (see Fig. 7.35) during the synchrotron experiments. The microstructure analysis at the end of the thermal cycle revealed the formation of wurtzite phase, nearly Ti-free fcc-TiN with an Al concentration of $x \approx 0.04$, as estimated from Eq. (7.1), and nearly Ti-free fcc-AlN.

After the thermal treatment the multilayer architecture is still recognizable in the DF STEM image (Fig. 7.36a) but considerable changes in comparison to the as-deposited state (Fig. 7.36b) are visible. For example the layer labeled as "A" in the as-deposited state which indicated the Ru enriched regions is not visible after the thermal treatment. Instead bright particles with a size of less than 10 nm are visible in the DF STEM image after thermal treatment which were identified as Ru by EDS analysis.

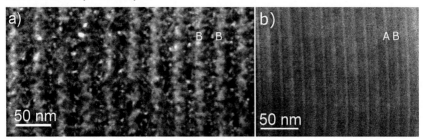

Fig. 7.36: DF STEM images of the cross section of the Ti-Al-N / Al-Ti-Ru-N multilayer coating of series III deposited at $U_B = -80$ V after thermal treatment at 950°C (a) and in the as-deposited state (b). The growth direction is from the right to the left of the images. The letters "A" and "B" indicate the Ru-rich and Ti-rich layers, respectively.

Furthermore, the contrast of the STEM DF image within the individual layers is not anymore constant as in the as-deposited state. This indicates that the concentration of the metallic ratios within the individual layers changes in the direction parallel to the layers. This was proven by an EDS line scan which was done perpendicular to the layer stacking in the Al-rich layer lying in between two layers labelled as "B". The differences in the composition along that line scan with a length of 150 nm are shown in Fig. 7.37a. Since the Ru content was too low to determine the Ru concentration, the intensity of the Ru Kα line is plotted in Fig. 7.37a. Although qualitative phase analysis of the X-ray diffraction patterns done after thermal treatment revealed the presence of w-AlN, nearly Ti-free fcc-AlN and nearly Al-free fcc-TiN (see Chapter 7.2.7.1), the determined [Al] / Σ[Me] ratio along the EDS line scan in the Al-rich layer never reached unity and ranged between 0.47 and 0.84. The average [Al] / Σ[Me] ratio was 0.63. It is assumed that in local regions within the TEM foil thickness the maximum [Al] / Σ[Me] ratio can be higher than the measured value of 0.84. The TEM foil thickness (see Fig. 7.37b) is a multiple of the crystallite size which is below 10 nm. Thus, several Al-rich and Ti-rich crystallites can be present within the whole TEM foil thickness. Due to this, the composition measured over the whole 50 nm to 80 nm thick TEM foil (see Fig. 7.37b) gives an average of several measured Al-rich and Ti-rich crystallites.

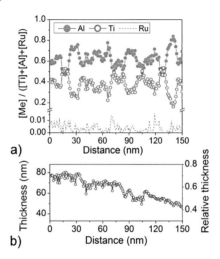

Fig. 7.37: Change of the [Me] / ([Ti]+[Al]+[Ru]) ratio within the Al-rich layer in the Ti-Al-N /Al-Ti-Ru-N multilayer of coating series III deposited at $U_B = -80$ V as determined by EDS analysis. The blue broken line in (a) shows the intensity of the Ru Kα line in arbitrary units. The TEM foil thickness as determined from the EELS zero loss signal is shown in (b) as absolute thickness (see Eq. (6.63)) and as relative thickness (see Eq. (6.60)) which is given in relative units of the mean free path for inelastic scattering.

Furthermore, the spatial resolution of the EDS measurement is limited by the broadening of the electron beam due to elastic scattering (see Chapter 6.3.3). According to Eq. (6.57) b_{ebeam} is in the range of ~ 4 nm for a 80 nm thick TEM foil made of (Ti,Al)N.

In order to reveal further information about the phase composition within the Al-rich layer, an EELS line scan, which recorded the energy loss at the N K edge, was performed simultaneously to the EDS line scan whose results are shown in Fig. 7.37. Although the measured EEL spectrum in each point along the EELS line scan represents an overlap of the EEL spectra of the phases present across the foil thickness, special features were found in the measured near edge structure of the nitrogen K edge that allows to differentiate between w-AlN and fcc-AlN and between w-AlN and fcc-TiN. The characteristic features of the near edge structure of the N K edge of fcc-AlN, w-AlN and fcc-TiN, as measured in reference samples, are shown in Fig. 6.7 in Section 6.3.3.2. Since fcc-Ti$_{1-x}$Al$_x$N would produce a near edge structure of the N K edge that is similar to the weighted sum of the binary spectra of fcc-TiN and fcc-AlN, as shown in Refs. [167, 176], it is not possible to differentiate between one crystal composed of fcc-Ti$_{1-x}$Al$_x$N and, e.g., two overlapping crystals composed of fcc-AlN and fcc-TiN with an identical averaged Al concentration x on the basis of the N K edge. Due to that, the following spectra, which were measured a long a line scan, were interpretated under the assumption that the phases present across the TEM foil thickness are fcc-AlN, fcc-TiN and w-AlN as indicated by the XRD analysis (see Section 7.2.7.2).

Four EEL spectra measured at the distances 17 nm, 68 nm, 90 nm and 140 nm in the Al-rich layer of the thermally treated Ti-Al-N / Al-Ti-Ru-N multilayer coating are shown in Fig. 7.38. The comparison of the near edge structure of the N K edge in these measured spectra with the near edge structure of the N K edge shown in the reference EEL spectra of fcc-AlN, w-AlN and fcc-TiN (see Fig. 6.7a, Fig. 6.7b and Fig. 6.7c) indicated the presence of these phases with a different volume fraction at the distances of 17 nm, 68 nm, 90 nm and 140 nm along the line scan measured within an Al-rich layer in the thermally treated sample (Fig. 7.38). This can be described at follows:

The Ti L$_{2,3}$ absorption edge is visible in all four spectra, but the intensity of the Ti L$_{2,3}$ absorption edge decreases from the position of 17 nm to the position of 140 nm which indicates a decreasing Ti content. This trend is also verified by EDS analysis (see Fig. 7.37) which showed also a decreasing [Ti] / ([Ti] + [Al]) ratio from 0.51 (17 nm) to 0.31 (68 nm) to 0.29 (90 nm) to 0.18 (140 nm).

The spectrum in Fig. 7.38a obtained at the distance of 17 nm from the start of the line scan is dominated by the presence of fcc-TiN which is indicated by the feature "A$_{fcc-TiN}$" and the increased intensity of the Ti L$_{2,3}$ edge. The second feature visible in the spectrum that is labelled by "B$_{w-AlN}$ / A$_{fcc-AlN}$" can be attributed to w-AlN or fcc-AlN or both.

The features "A$_{w-AlN}$" and "B$_{w-AlN}$" visible in the spectrum in Fig. 7.38b, which was obtained at the position of 68 nm, indicated the dominance of w-AlN at this position. Still, the feature labelled "A$_{fcc-TiN}$" and the Ti L$_{2,3}$ edge are visible, but with a lower intensity as compared to the spectrum in Fig. 7.38a. This indicated the existence of fcc-TiN with a lower volume fraction which is also expected from the lower [Ti] / ([Ti] + [Al]) ratio measured by EDS.

The spectrum measured at the distance of 90 nm from the start of the line scan (see Fig. 7.38c) contains the features typical for fcc-AlN. These features are labelled by "A$_{fcc-AlN}$" and "D$_{fcc-AlN}$".

Fcc-AlN is the dominating phase at this position. Additionally, the feature "$A_{fcc-TiN}$" and the Ti $L_{2,3}$ edge imply the presence of the fcc-TiN phase with a low volume fraction.

The spectrum obtained at the distance of 140 nm after the start of the line scan (Fig. 7.38d) is similar to the spectrum at 90 nm. It contains also the features typical for fcc-AlN, namely "$A_{fcc-AlN}$" and "$D_{fcc-AlN}$". Like in the spectrum at 90 nm, the feature "$A_{fcc-TiN}$" and the Ti $L_{2,3}$ edge are visible, but both intensities are lower as compared to the spectrum shown in Fig. 7.38c. Thus, at this position the lowest fraction of fcc-TiN is expected which is facilitated by the lowest [Ti] / ([Ti] + [Al]) ratio as measured by EDS among the four positions.

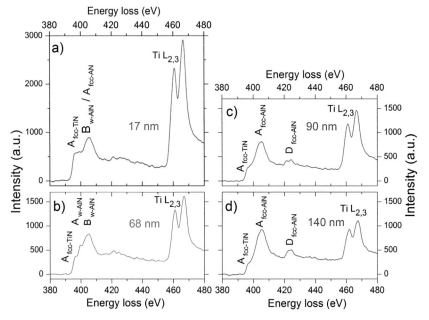

Fig. 7.38: Selected EEL spectra that were measured along the EELS line scan in an Al-rich layer of the thermally treated Ti-Al-N / Al-Ti-Ru-N multilayer of coating series III deposited at U_B = -80 V showing the N K edge and the Ti $L_{2,3}$ edge. In each spectrum a certain phase is prevailing: (a) fcc-TiN, (b) w-AlN, (c,d) fcc-AN. Significant features are marked by A, B, D.

The variation of the metallic ratio across the multilayer stacking of the Ti-Al-N / Al-Ti-Ru-N multilayer coating of series III which was deposited at U_B = -80 V is shown Fig. 7.39 after thermal treatment and in the as-deposited state. According to the EDS line scans shown in Fig. 7.39b and Fig. 7.39f the medium [Al] / Σ[Me] ratio averaged over at least two bi-layer periods is 0.43 after thermal treatment and 0.47 in the as-deposited state and the [Ti] / Σ[Me] ratio is 0.56 and 0.52, respectively. In contrast to these results, EPMA yielded an average [Al] / Σ[Me]

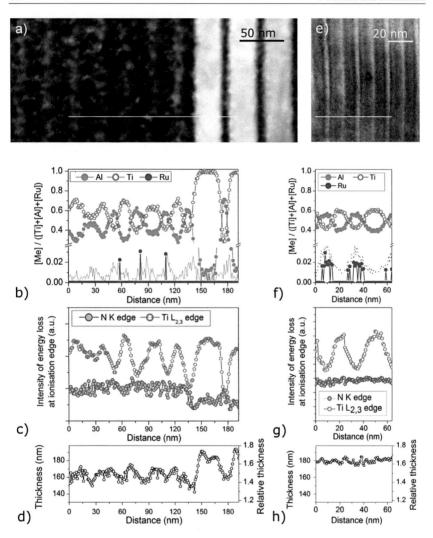

Fig. 7.39: DF STEM images of the Ti-Al-N / Al-Ti-Ru-N multilayers of coating series III deposited at U_B = -80 V after thermal treatment at 950°C (a) and in as-deposited state (e). The white lines show the position of the analytical TEM line scans. The results of the EDS (b,f) and EELS (c,g) analysis are given below the DF STEM image for the respective coating. The blue dotted lines in figures (b, f) show the intensity of the Ru Kα line which was observed using EDS analysis. The TEM foil thickness as determined from the low-loss spectrum is shown in figures (d, h) as absolute thickness and as relative thickness given in relative units of the mean free path for inelastic scattering.

ratio of 0.53 (see Table 7.1). This shows that the [Al] / Σ[Me] ratio determined by EDS in TEM is underestimated. This can be attributed to the used standardless quantification method of the EDS signals in TEM and emphasizes the fact that the EDS results should reveal primarily differences in the composition along a line scan rather than to determine the absolute composition (see also Chapter 6.3.3). However, the multilayer architecture due to the variation of the metallic ratios across the multilayer is recognizable from the EDS line scan obtained after thermal treatment (see Fig. 7.39b). The bi-layer period of ~ 29 nm can be identified from it. It is similar like in the as deposited state. Although XRD revealed nearly Al-free fcc-TiN, w-AlN and nearly Ti-free fcc-AlN after thermal treatment, the maximum Ti-ratio in the Ti-rich layer as well as the maximum Al ratio in the Al-rich layer were significantly below 1. This is attributed to the fact that the thickness of the TEM foil is in the range of ~ 160 nm which is a multiple of the crystallite size as explained above on page 121. The absolute thickness and the relative thickness of the TEM foil are given in Fig. 7.39d and Fig. 7.39h for the TEM samples showing the state after and before the thermal treatment, respectively. Both values were determined from the intensity of the EEL spectra related to the intensity of the zero-loss peak in the EEL spectra according to Eq. (6.63) and Eq. (6.60) assuming a medium Z_{eff} = 13.41 (see Section 6.3.3.2). Moreover, significant beam broadening (cf. Fig. 6.4) is expected due to the thick TEM foil which limits the spatial resolution of the EDS analysis.

The variation of the Ti concentration across the multilayers was also proven by EELS analysis of the Ti $L_{2,3}$ edge in the as-deposited state and as well as after thermal treatment. This variation is visible from the intensity of the energy loss at the Ti $L_{2,3}$ edge as shown in Fig. 7.39c and Fig. 7.39f for the thermally treated and virgin coating, respectively. Additionally, the intensity of the energy loss at the nitrogen K ionization edge is plotted in both figures. The intensity of the respective ionization edges were determined after the removal of plural scattering effects from the EEL spectra. Thus, the variation of the intensity of the nitrogen K edge for the Ti-Al-N / Al-Ti-Ru-N multilayer coating (U_B = -80 V) after thermal treatment at 950 °C (see Fig. 7.39c) is not attributed to the TEM foil thickness. In Ref. [166] it could be shown that the intensity of the nitrogen K edge in coatings based on Ti-Al-N is related to the present phases (fcc-TiN, fcc-Ti$_{1-x}$Al$_x$N and fcc-AlN / w-AlN). According to Ref. [166] the intensity of the nitrogen K edge measured in fcc-AlN and w-AlN rises by a factor of 2.4 - 2.6 as compared to fcc-TiN and the intensity of the nitrogen K edge measured in fcc-Ti$_{1-x}$Al$_x$N ($0.4 \leq x \leq 0.65$) increases by a factor of 1.4 – 1.5 in comparison to fcc-TiN (see Section 6.3.3.2 page 56). These intensity ratios of the nitrogen K ionization edge explain the drop of the intensity of the nitrogen K edge that was measured along the EELS line scan of the thermally treated coating in Fig.

7.39a in the range of 148 to 170 nm. In this region nearly Al-free fcc-TiN is present (see Fig. 7.39b) whereas in the range of 0 to 140 nm a mixture of nearly Al-free fcc-TiN, w-AlN and nearly Ti-free fcc-AlN is present.

The analysis of the near edge structure of the nitrogen K edge revealed further information about the phases present after thermal treatment at 950 °C. As in the case of the EELS line scan within the Al-rich layer (see above Fig. 7.37 and Fig. 7.38) significant features in the near edge structure of the N K edge could be identified that indicate the presence of w-AlN, fcc-AlN and fcc-TiN although the measured EEL spectrum of each measuring point along the EELS line scan represents an overlap of the EEL spectra of several phases present across the foil thickness.

7.2.7.3   Interfaces between fcc and wurtzite phase after thermal treatment

In order to determine possible orientation relationships between w-AlN and fcc-(Ti,Al)N present after the annealing sequence shown in Fig. 6.2b, HRTEM of the Ti-Al-N / Al-Ti-Ru-N multilayer coating of series III deposited at $U_B = -80$ V was done in an area close to the hole of the TEM foil which was characterized by a low thickness. In fact the same orientation relationship between fcc-(Ti,Al)N and w-AlN that was previously observed in the as deposited $Ti_{0.44}Al_{0.56}N$ coating ($U_B = -40$ V) (see Fig. 7.7) and in the as deposited Ti-Al-N / Al-Ti-N multilayer coating of series I ($U_B = -40$ V) (see Fig. 7.26) could be identified in the HRTEM image shown in Fig. 7.40. The FFT of the area marked by a yellow square in the upper left part of the HRTEM image shows the fcc-(Ti,Al)N phase with the $[001]_{fcc}$ direction being parallel to the normal of the TEM foil plane (see Fig. 7.40c). The indices of the reflections correspond to the indices of the simulated electron diffraction pattern of the fcc phase (black spots) in Fig. 7.40g. The FFT of the adjacent area marked by an orange square in the HRTEM image revealed additionally to the fcc-(Ti,Al)N phase the wurtzite phase (see Fig. 7.40d) with the $[001]_w$ direction being parallel to the normal of the plane. For a better visibility, the reflections of the w-AlN phase in the FFT in Fig. 7.40d are highlighted in Fig. 7.40e and can be indexed by the simulated electron diffraction pattern of w-AlN (orange spots) given in Fig. 7.40g. A higher magnification of the fcc-(Ti,Al)N / w-AlN interface is shown in the HRTEM image in Fig. 7.40b and its corresponding FFT in Fig. 7.40f. The FFT corresponds exactly to the simulation of the diffraction patterns of both phases with the directions $[001]_{fcc}$ and $[001]_w$ being parallel to the incident beam direction and the $(\bar{2}10)_w$ plane being parallel to the $(\bar{1}\bar{1}0)_{fcc}$ plane. A schematic presentation of this orientation relationship is illustrated in Fig. 7.8 and Fig. 7.9.

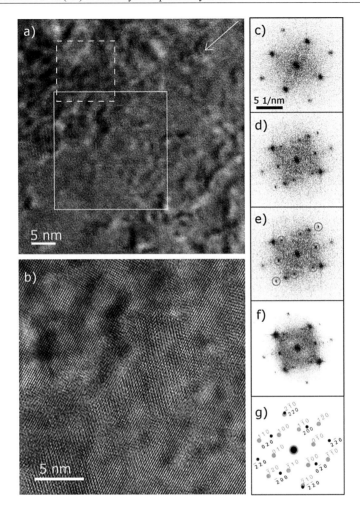

Fig. 7.40: HRTEM images (a,b) and FFTs (c-f) of the Ti-Al-N / Al-Ti-Ru-N multilayer coating of series III deposited at $U_B = -80$ V after thermal treatment at 950 °C. The white arrow in fig. (a) indicates the growth direction of the coating. HRTEM of the area marked by the white square was done at a higher magnification and is shown in fig. (b). The FFT in fig. (c) corresponds to the yellow marked square in fig. (a) and shows the fcc-(Ti,Al)N phase with the $(001)_{fcc}$ plane parallel to the TEM foil plane. The FFT in fig. (d) was done in the region marked by the orange square and reveals the wurtzite phase with the $(001)_w$ plane parallel to the TEM foil plane. The reflections of wurtzite phase are highlighted in figure (e). The FFT in fig. (f) is from the whole HRTEM image shown in fig. (b) and illustrates the OR of the fcc and wurtzite phase. The simulated electron diffraction pattern with $(\bar{1}10)_{fcc} \parallel (\bar{2}10)_w$ and $[001]_{fcc} \parallel [001]_w$ is given in fig. (g) where the black and orange marked diffraction spots correspond to the fcc and wurtzite phase, respectively.

SAED done in an area with a diameter of ~ 100 nm which included the region shown in the HRTEM image in Fig. 7.40a proved the presence of this kind of orientation relationship, see Fig. 7.41a. In Fig. 7.41a the diffraction spots of fcc-(Ti,Al)N are labeled by the black indices. The direction of the incident beam corresponds to [001]$_{fcc}$. The diffraction spots originating from the w-AlN phase oriented with the [001]$_w$ direction parallel to surface normal and the $(\bar{2}10)_w$ plane being parallel to the $(\bar{1}\bar{1}0)_{fcc}$ plane are indicated by the orange arrows.

However, w-AlN does not form with this relationship (($\bar{1}\bar{1}0)_{fcc}$ ∥ $(\bar{2}10)_w$ and [001]$_{fcc}$ ∥ [001]$_w$) in the entire sample after the used annealing cycle. This can be seen from the SAED from a larger sample area with a diameter of ~ 540 nm shown in Fig. 7.41b. In the SAED a preferred orientation of the fcc-(Ti,Al)N phase is apparent since the intensity along the diffraction rings is not evenly distributed. Rather discrete diffraction spots are visible for the

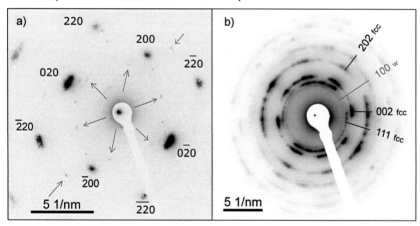

Fig. 7.41: SAED of the Ti-Al-N / Al-Ti-Ru-N multilayer coating of series III deposited at U$_B$ = -80 V after thermal treatment at 950°C. The diameter of the region contributing to the SAED pattern was ~ 100 nm in figure (a) and ~540 nm in figure (b). The orange arrows in figure (a) indicate the reflections of wurtzite phase that were observed in the HRTEM image of Fig. 7.40. The black labelled indices correspond to fcc-(Ti,Al)N.

fcc-(Ti,Al)N phase, e.g. (002)$_{fcc}$. This is a residue from the growth texture formed during deposition as shown in Section 7.2.6. In contrast to the fcc-(Ti,Al)N phase, no discrete diffraction spots are visible e.g. for the (100)$_w$ reflections of w-AlN. This can be attributed to different reasons: (i) The intensity of the w-AlN diffraction spots is much lower than the ones of fcc-(Ti,Al)N so that intensity variations along a certain diffraction ring are hard to identify. (ii) Other orientation relationships between w-AlN and fcc-(Ti,Al)N which were found experimentally e.g. [1$\bar{1}$0]$_{fcc}$ ∥ [100]$_w$ and ($\bar{1}11$)$_{fcc}$ ∥ (002)$_w$ [48, 226] or [0$\bar{1}$0]$_{fcc}$ ∥ [0$\bar{1}\bar{1}$]$_w$ and

$(002)_{fcc}$ ∥ $(01\bar{1})_w$ [77] are possible. (iii) TEM was done after the whole thermal cycle (see Fig. 6.2b) when a considerable amount of w-AlN formed. The estimated molar fraction is $\sim (20 \pm 5)$ mol %. Hence, a substantial fraction of the formed w-AlN could be oriented randomly.

Nevertheless, due to the fact that the above described orientation relationship between fcc-(Ti,Al)N and w-AlN ($[001]_{fcc}$ ∥ $[001]_w$ and $(\bar{2}10)_w$ ∥ $(\bar{1}\bar{1}0)_{fcc}$) was found coincidentally in three different Ti-Al-N based coatings and since the lattice misfit especially in the $[\bar{1}\bar{1}0]_{fcc}$ and $[\bar{1}00]_w$ direction is low (see Chapter 7.1.5), it is assumed that this orientation relationship constitutes a favored one between both phase. In order to study it in more detail, w-AlN was grown heteroepitaxially on fcc-TiN by magnetron sputtering and the orientation relationships between both phases were studied by electron diffraction using TEM, which is presented in the next Chapter 7.3.

7.3   Heteroepitaxial TiN / AlN / TiN layers deposited by MS

Several orientation relationships between the wurtzite and fcc phase in Ti-Al-N based coatings were found experimentally in literature and are summarized in Table 7.3.

Table 7.3: Summary of published orientation relationships between fcc-(Ti,Al)N and w-AlN found experimentally in Ti-Al-N based coatings.

Reference	Orientation relationship	Label
Ref. [46]	$[001]_{fcc}$ ‖ $[001]_w$ and $(1\bar{1}0)_{fcc}$ ‖ $(\bar{1}10)_w$	
Ref. [38]	$[001]_{fcc}$ ‖ $[\bar{1}\bar{1}0]_w$ and $(1\bar{1}0)_{fcc}$ ‖ $(\bar{1}10)_w$	
Refs. [48, 226]	$[11\bar{1}]_{fcc}$ ‖ $[001]_w$ and $(0\bar{1}\bar{1})_{fcc}$ ‖ $(110)_w$	
Ref. [77]	$[0\bar{1}0]_{fcc}$ ‖ $[0\bar{1}\bar{1}]_w$ and $(001)_{fcc}$ ‖ $(01\bar{1})_w$	
Ref. [202, 203]	$[001]_{fcc}$ ‖ $[111]_w$ and $(1\bar{1}0)_{fcc}$ ‖ $(\bar{1}01)_w$	
Ref. [202, 203]	$[001]_{fcc}$ ‖ $[001]_w$ and $(\bar{1}10)_{fcc}$ ‖ $(\bar{2}10)_w$ and $(1\bar{1}0)_{fcc}$ ‖ $(0\bar{1}0)_w$	Eq. (7.19)

During the analysis of the local orientation relationship (OR) between wurtzite and fcc phase in the Ti-Al-N monolayer and Ti-Al-N / Al-Ti-(Ru)-N multilayer coatings, the OR characterized by $(\bar{1}10)_{fcc}$ ‖ $(\bar{2}10)_w$ and $[001]_{fcc}$ ‖ $[001]_w$ (see Eq. (7.19) in Table 7.3) that was previously identified in Ref. [202] emerged several times, namely in the $Ti_{0.44}Al_{0.56}N$ coating (see Section 7.1.5, Fig. 7.7), in the Ti-Al-N / Al-Ti-N multilayer coating (series I) deposited at $U_B = -40$ V (see Section 7.2.3, Fig. 7.26) as well as in the thermally treated Ti-Al-N / Al-Ti-Ru-N multilayer coating (series III) deposited at $U_B = -80$ V (see Section 7.2.7.2, Fig. 7.40).

As mentioned in Section 7.1.5, this OR is characterized by a moderate misfit of ~ 3.6 % along the $[\bar{1}10]_{fcc}$ and $[\bar{1}00]_w$ directions. On the contrary, a huge misfit of ~ 16 % along the $[001]_{fcc}$ and $[001]_w$ directions would be present if a flat interface between both phases is assumed to be parallel to the $(1\bar{1}0)_{fcc}$ plane or $(0\bar{1}0)_w$ plane as shown in Fig. 7.9 (see page 76).

In order to study the interface containing the $[001]_{fcc}$ and $[001]_w$ directions in more details, wurtzitic AlN was deposited onto a $(1\bar{1}0)_{fcc}$ oriented fcc-TiN layer using magnetron sputtering. Additionally, w-AlN was grown on top of a $(00\bar{1})_{fcc}$ oriented fcc-TiN film. Inspired by the work of Adibi *et al.* [70], commercialized fcc-MgO single crystal substrates being $(1\bar{1}0)$ and $(00\bar{1})$ oriented were chosen in order to grow $(1\bar{1}0)_{fcc}$ as well as $(00\bar{1})_{fcc}$ oriented fcc-TiN films by magnetron sputter deposition. Fcc-TiN and fcc-MgO have the same crystal structure (space group: Fm$\bar{3}$m) and similar lattice parameters ($a_{\text{TiN}} = 0.42418$ nm, $a_{\text{MgO}} = 0.421$ nm) which results in a low misfit (f_4) that is according to Eq. (7.20) smaller than 1 %.

$$f_4 = 2 \cdot \frac{a_{\text{TiN}} - a_{\text{MgO}}}{a_{\text{TiN}} + a_{\text{MgO}}} \qquad\qquad (7.20)$$

Thus, the growth of fcc-TiN with the same orientation as the fcc-MgO single crystal substrate is expected.

The details of the deposition process and the analysis of the heteroepitaxy between fcc-MgO and fcc-TiN as well as fcc-TiN and w-AlN are described in the following Sections. The results were partially published in Refs. [203, 214].

7.3.1   Preparation of TiN / AlN / TiN layer stacks

In order to investigate the orientation relationship (OR) between fcc-TiN and w-AlN in detail, a layer stack composed of fcc-TiN / w-AlN / fcc-TiN with an individual layer thickness of ~ 110 nm was deposited onto ($1\bar{1}0$) and ($00\bar{1}$) oriented fcc-MgO (space group: Fm$\bar{3}$m) single crystals by reactive magnetron sputtering (MS) using a Balzers PLS 500 deposition chamber. The fcc-MgO single crystals were purchased from CrysTec GmbH as epitaxy-ready (epi) polished substrates with a size of $20 \times 20 \times 0.5$ mm^3.

The TiN and AlN layers were produced by reactive MS in a mixed argon / nitrogen atmosphere using a pure titanium and a pure aluminium target with a diameter of 90 mm. The targets were installed on two circular planar magnetrons. The normal of the substrate holder and the normal of the targets were antiparallel. The three layers (TiN, AlN and TiN) were deposited successively by the rotation of the substrate to face the respective target. Prior the deposition, the hysteresis curves (cathode voltage as function of the nitrogen flow and pressure as function of the nitrogen flow [227]) for reactive sputter deposition of TiN and AlN were recorded in order to determine the appropriate N flows. After selecting the individual gas flows, the sputter rates were determined. The individual deposition parameters for the TiN and AlN layer with regard to the sputter rate and Ar / N flow are given in Table 7.4.

Table 7.4: Deposition parameters of the individual TiN and AlN layers deposited by MS.

Layer	P_{target} (kW)	Ar flow (sccm)	N flow (sccm)	Deposition time (s)
TiN	0.7	32	3.7	90
AlN	0.25	29	7	480

The vacuum chamber was pumped to a base pressure of $2.5 \cdot 10^{-6}$ mbar. The argon and nitrogen gases passed gas cleaners before feeding them into the deposition chamber. The deposition of the fcc-TiN / w-AlN / fcc-TiN multilayer stack was done at 400°C and the substrate holder was biased to $U_B = -80$ V.

For the heteroepitaxy studies, the first TiN layer (called TiN seed layer) that was deposited onto the fcc-MgO substrate and the adjacent w-AlN layer were of the main interest. The second TiN layer was deposited as kind of protection layer at the end of the process.

Schematic drawings of the fcc-TiN / w-AlN / fcc-TiN layer stacks deposited onto the $(1\bar{1}0)$ and $(00\bar{1})$ oriented fcc-MgO single crystals are given in Fig. 7.42. For the analysis of the fcc-TiN / w-AlN interface using SAED and HRTEM / FFT four TEM cross section samples with certain orientations with respect to the fcc-MgO crystal were prepared by FIB (see Fig. 7.42). The samples FIB_1 and FIB_2 were prepared from the fcc-TiN / w-AlN / fcc-TiN layer stack that was deposited onto the $(1\bar{1}0)$ oriented fcc-MgO single crystal. The samples FIB_3 and FIB_4 were prepared from the fcc-TiN / w-AlN / fcc-TiN layer stack that was deposited onto the $(00\bar{1})$ oriented fcc-MgO single crystal. The positions of the TEM cross section samples

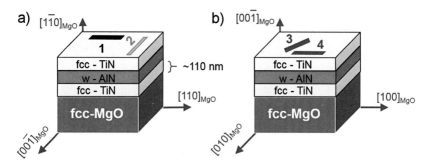

Fig. 7.42: Schematic drawing of the fcc-TiN / w-AlN / fcc-TiN layer stack deposited onto $(1\bar{1}0)$ (a) and $(00\bar{1})$ (b) oriented fcc-MgO single crystal substrates showing the position of the FIB cross section samples (FIB_1 to FIB_4).

with respect to the fcc-MgO crystal are given schematically in Fig. 7.43. With the samples FIB_1 and FIB_4 it was intended to check the OR between w-AlN and fcc-TiN for the $\langle 001 \rangle_{fcc}$ zone axis which corresponds to the zone axis $[001]_{fcc}$ that was used during the HRTEM / FFT analysis of the $Ti_{0.44}Al_{0.56}N$ monolayer and the Ti-Al-N / Al-Ti-(Ru)-N multilayers revealing the OR given by Eq. (7.19) (or cf. Fig. 7.8). The intention of the sample FIB_2 was to investigate the TiN / AlN interface parallel to the $[00\bar{1}]_{fcc}$ direction for the zone axis $\langle \bar{1}10 \rangle_{fcc}$ as shown in Fig. 7.9 for the expected OR given in Eq. (7.19). The sample FIB_3 should illuminate the TiN / AlN interface for the same zone axis as the sample FIB_2 but with the TiN / AlN interface parallel to the $[1\overline{1}0]_{fcc}$ direction.

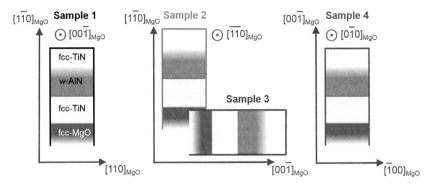

Fig. 7.43: Schematic drawing of the TEM cross section samples showing their position with respect to the orientation of the fcc-MgO single crystal. The direction pointing out of the plane of the paper corresponds to electron beam direction with respect to the fcc-MgO substrate. The size of the individual layers is not to scale.

7.3.2 Internal interfaces studied by heteroepitaxial growth

TEM bright field images of the cross sections of the FIB samples of the TiN / AlN / TiN layer stacks grown on $(1\overline{1}0)_{fcc}$ and on $(00\overline{1})_{fcc}$ oriented fcc-MgO are given in Fig. 7.44 and Fig. 7.45, respectively. The thickness of each layer was in the range of ~ 110 to ~ 117 nm.

SAED done in an area with a diameter of ~ 97 nm could confirm that the first fcc-TiN layer was grown heteroepitaxially with the same orientation as the used fcc-MgO single crystal substrates. This is illustrated by the SAED patterns of the substrate and the adjacent fcc-TiN layer shown at the bottom of Fig. 7.44 and Fig. 7.45. Furthermore, the SAED patterns showed that the zone axis of the samples FIB_1 and FIB_2 is parallel to the $[00\overline{1}]$ and $[\overline{1}\overline{1}0]$ direction of the fcc-MgO substrate, respectively. The zone axis of the samples FIB_3 and FIB_4 is parallel to the $[\overline{1}\overline{1}0]$ and $[0\overline{1}0]$ direction of the fcc-MgO substrate, respectively.

The heteroepitaxial growth of the fcc-TiN seed layer on the fcc-MgO substrate was additionally investigated by RSMs, which were measured by HRXRD and which are shown in the Appendix (see Fig. A. 6 and Fig. A. 8). The RSMs indicated a strained TiN layer. Furthermore, the broadening of the reciprocal lattice points of the fcc-TiN layer suggested a low dislocation density in the fcc-TiN seed layer (further details are given in the Appendix on page 194).

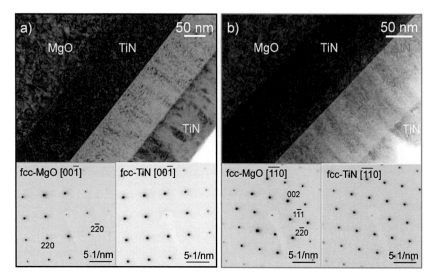

Fig. 7.44: TEM BF image of the TiN / AlN / TiN layer stack deposited onto (1$\bar{1}$0) oriented fcc-MgO substrate of the samples FIB_1 (a) and FIB_2 (b). The SAED patterns shown at the bottom of the figure were measured in the fcc-MgO substrate and in the adjacent fcc-TiN layer.

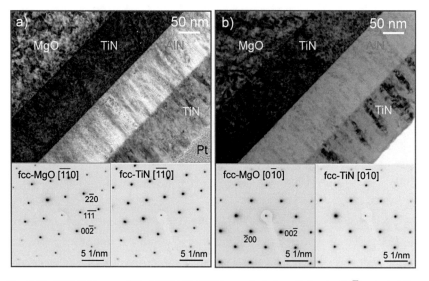

Fig. 7.45: TEM BF image of the TiN / AlN / TiN layer stack deposited onto (00$\bar{1}$) oriented fcc-MgO substrate of the samples FIB_3 (a) and FIB_4 (b). The SAED patterns shown at the bottom of the figure were measured in the fcc-MgO substrate and in the adjacent fcc-TiN layer.

Regarding the TEM BF images (Fig. 7.44 and Fig. 7.45) that were done using diffraction contrast under multiple beam conditions, contrasts in the single crystalline MgO substrate and in the adjacent fcc-TiN layer are apparent. Neither differences in the orientation (see Appendix page 194) nor in the composition are expected within the substrate or within the TiN seed layer which could be responsible for these contrasts. Possibly, these contrasts may arise from a corrugated surface caused by etching effects during TEM foil preparation. However, the explanation of the origin of such contrasts was not the aim of this work.

Columnar grains are visible in the AlN layer and in the 2nd TiN, layer whereas no columnar grains are visible in the 1st TiN layer. SAED of the 2nd TiN layer showed the polycrystalline character of this layer. However, the investigations within this study are focused on the orientation relationship between the 1st TiN layer and the neighbouring w-AlN layer and will be discussed at first for the layer stack grown on the $(1\bar{1}0)_{fcc}$ and then for the layer stack grown on the $(00\bar{1})_{fcc}$ oriented fcc-MgO substrate.

7.3.2.1 Wurtzite AlN grown on $(1\bar{1}0)_{fcc}$ oriented fcc-TiN

SAED done in the AlN layer of the sample **FIB_2** revealed the electron diffraction pattern given in Fig. 7.46a. The diameter of the analysed region was ~ 97 nm and covered nearly the whole layer thickness of ~ 113 nm. The diffraction spots of the fcc-TiN seed layer are not included in the SAED pattern, but their positions are indicated by the white lattice and are labelled by white indices in Fig. 7.46a. The white arrows in Fig. 7.46a indicate the orientation of the fcc-MgO substrate which is identical with the orientation of the fcc-TiN seed layer. Hence, in both cases the zone axis is $[\bar{1}\bar{1}0]_{fcc}$. The interface between the fcc-TiN seed layer and the AlN layer is parallel to the $[001]_{MgO}$ direction.

The diffraction pattern shown in Fig. 7.46a can be indexed by the simulated diffraction pattern given in Fig. 7.46b. In order to obtain the indexing, the origin of the measured SAED (Fig. 7.46a) was illuminated by local orientation analysis done by HRTEM / FFT as shown in Fig. 7.47. HRTEM / FFT showed that the w-AlN layer is composed of columnar grains with a width of 10 to 20 nm. FFT done in an area of 23 nm × 23 nm (see orange square), which contained apparently two adjacent grains, led to the diffraction pattern shown in Fig. 7.47c. This pattern is similar to the electron diffraction pattern observed by SAED (see Fig. 7.46a). FFT obtained from an area of approx. 11 nm × 11 nm and performed separately in each adjacent grain (see

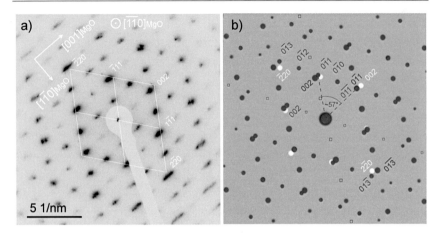

Fig. 7.46: SAED pattern of the AlN layer in the sample FIB_2 (a). The white arrows indicate the orientation with respect to the fcc-MgO substrate. The white lattice marks the diffraction pattern of the fcc-TiN seed layer whose diffraction spots are not included in the measured SAED. The white indices in (a) correspond to the fcc-TiN seed layer. The simulated electron diffraction pattern of w-AlN with two different orientations of w-AlN (w2.1 = red, w2.2 = blue) and the reduced diffraction pattern of fcc-TiN (white spots) (b). The $[\bar{1}10]_{fcc}$ and $[\bar{1}00]_{w2.1/w2.2}$ directions are perpendicular to the plane of the paper. The open squares correspond to systematically absent reflections due to space group (e.g. 001) which could be observed due to double diffraction.

red and blue squares in Fig. 7.47a) revealed the diffraction patterns shown in Fig. 7.47d and Fig. 7.47e.

Although the analysed regions lay apparently in separated grains, the diffraction patterns contain also the information of other columnar grains that are present in the TEM foil since its thickness of 40 to 50 nm is a multiple of the column width. The TEM foil thickness was estimated from the EELS spectrum taking into account the total intensity of the spectrum and the intensity of the zero-loss peak according to Eq. (6.63). However, in each pattern in Fig. 7.47d and Fig. 7.47e one certain orientation is dominating. Both orientations are characterized by the $[\bar{1}00]_w$ zone axis. The dominating orientation in Fig. 7.47e can be described by the red marked diffraction spots in the simulated diffraction pattern in Fig. 7.46b. The dominating orientation in the FFT in Fig. 7.47d can be obtained by a clockwise rotation of the diffraction pattern in Fig. 7.47e by an angle of approximately 60° around the zone axis yielding a similar diffraction pattern as indicated by the blue spots in the simulated electron diffraction pattern Fig. 7.46b.

It has to be emphasized that the red and blue marked diffraction spots in the simulated diffraction pattern in Fig. 7.46b show a special case of two orientations with the $[\bar{1}00]_w$ zone axis. Because the orientation "w2.2" indicated by the blue spots was obtained by the clockwise rotation of the w-AlN crystal "w2.1" (red spots) by an angle of 57° around the $[\bar{1}00]_{w2.1}$ zone axis. The two special ORs given in the simulated diffraction pattern are characterized by:

w2.1: $[\bar{1}\bar{1}0]_{fcc} \parallel [\bar{1}00]_{w2.1}$ and $(002)_{fcc} \parallel (0\bar{1}\bar{1})_{w2.1}$

w2.2: $[\bar{1}\bar{1}0]_{fcc} \parallel [\bar{1}00]_{w2.2}$ and $(002)_{fcc} \parallel (0\bar{1}1)_{w2.2}$

In this special case, the $\{101\}$ plane that is perpendicular to the interface, is in common for both orientations. The angle between the $(0\bar{1}1)_{w2.1}$ and $(0\bar{1}\bar{1})_{w2.1}$ planes of the crystallite group "w2.1" (see red spots in Fig. 7.46b) is 56.81° and agrees with the rotation angle used for the transformation of the crystal "w2.1" to the crystal "w2.2". After this rotation, in Fig. 7.46b the $(0\bar{1}1)_{w2.2}$ plane of the type "w2.2" takes the position of the $(0\bar{1}\bar{1})_{w2.1}$ plane of the type "w2.1". According to the theoretical calculations by Béré et al. [228, 229] the $\{101\}$ planes are beside the $\{102\}$ and $\{103\}$ planes possible twin boundaries in wurtzite structures. Furthermore, Horiuchi et al. [230] observed the $\{101\}$ twin boundary in w-AlN whiskers that were produced by sublimation. Thus, it is deduced that twins can be formed within the w-AlN layer. In the special case shown in the simulated electron diffraction pattern the twin boundary corresponds to the $(0\bar{1}\bar{1})_{w2.1}$ plane. This means that the twin boundaries would be perpendicular to the interface and could constitute the column boundaries.

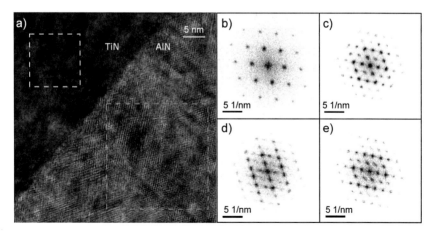

Fig. 7.47: HRTEM image of the TiN / AlN interface in the sample FIB_2 (a). The FFT in (b) originates from fcc-TiN marked by the white square. The FFT in (c) corresponds to w-AlN (orange square). The FFTs in (d) and (e) originate from the blue and red square, respectively.

The blue marked diffraction spots in the simulated diffraction pattern in Fig. 7.46b can be also obtained from the orientation "w2.1" by a $180°$ rotation around the $[1\overline{1}0]_{MgO}$ direction leading to the orientation labelled "w2.2a" which can be described as follows:

"w2.2a": $[\overline{1}\,\overline{1}0]_{fcc} \parallel [100]_{w2.2a}$ and $(002)_{fcc} \parallel (011)_{w2.2a}$

However, if the lattice parameters $a = 0.311$ nm and $c = 0.498$ nm [190] are assumed for w-AlN, the special orientation of "w2.1" and "w2.2" (or "w2.2a"), as shown in Fig. 7.46b and described above, implies that the lattice planes $(01\overline{3})_{w2.1}$ and $(0\overline{1}\overline{3})_{w2.2}$ (or $(01\overline{3})_{w2.2a}$) make an angle of $6.4°$ whereby the bisecting line is parallel to the $[1\overline{1}0]_{MgO}$ direction which is parallel to the normal of the interface. Instead of two discrete diffraction spots of $(01\overline{3})_{w2.1}$ and $(0\overline{1}\overline{3})_{w2.2}$ (or $(01\overline{3})_{w2.2a}$) separated by $6.4°$ as shown in the simulation in Fig. 7.46b, two broadened diffraction spots are visible in the measured SAED pattern in Fig. 7.46a. This indicates that further orientations close to the ones described by "w2.1" and "w2.2" (or "w2.2a") are present. This is also supported by the RSMs with $\vec{q}_x \parallel [00\overline{1}]_{MgO}$ and $\vec{q}_z \parallel [1\overline{1}0]_{MgO}$ that are shown in

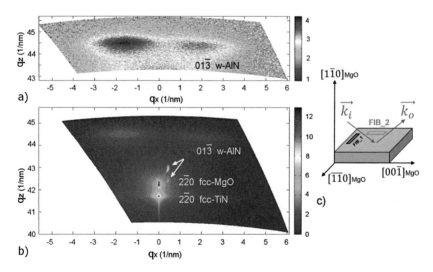

Fig. 7.48: Reciprocal space map of the TiN / AlN / TiN layer stack deposited onto the $(1\overline{1}0)$ oriented fcc-MgO single crystal showing the positions and shape of the reciprocal lattice points corresponding to the $\{103\}_w$ planes of the w-AlN crystals (a) and the $\{110\}_{fcc}$ planes of the fcc-MgO substrate and the $\{110\}_{fcc}$ planes of the fcc-TiN seed layer (b). The features marked by the white arrows in figure (b) indicate an artefact. The colour scale refers to the intensity given as natural logarithm. The orientation of the sample with respect to the incident $\vec{k}_i$ and outgoing $\vec{k}_o$ wave vector during the RSM measurement is displayed in figure (c) which corresponds to $\vec{q}_z \parallel [1\overline{1}0]_{MgO}$ and $\vec{q}_x \parallel [00\overline{1}]_{MgO}$.

Fig. 7.48 and which were obtained from a set of XRD rocking curve measurements using the D8 ADVANCE diffractometer from Bruker AXS (see Section 6.2.4). In the RSM given in Fig. 7.48b the reciprocal lattice points attributed to the $(2\overline{2}0)_{MgO}$ and $(2\overline{2}0)_{TiN}$ planes, which are parallel to the surface of the TiN / AlN / TiN layer stack, are visible at $q_x = 0$ and $q_z = 42.2$ nm^{-1} (fcc-TiN) and 41.7 nm^{-1} (fcc-MgO). Close to the reciprocal lattice points $(2\overline{2}0)_{MgO}$ and $(2\overline{2}0)_{TiN}$ two spots of increased intensity appear in the RSM shown in Fig. 7.48b. They are considered as artefacts of the measurement as discussed in the Appendix (see page 197). Additionally, the reciprocal lattice points attributed to the $\{10\overline{3}\}_w$ planes are indicated by light blue shading in the q_x-range of approx. -3 to 3 nm^{-1} and at $q_z \approx 44.5$ nm^{-1} in Fig. 7.48. They are better visible in Fig. 7.48a where two broadened reciprocal lattice points are visible. The rocking curve, measured at $2\theta_B = 66°$ for the 013 diffraction line of w-AlN (see Fig. 7.49), shows that the maxima of both lattice points are separated by $\Delta\omega = 4.4°$. The same was observed for the rocking curve measurement performed on the HR diffractometer from Seifert FPM without using any detector slit (see Appendix Fig. A. 7). On the one hand, this deviation from the expected $\Delta\omega$ of 6.4° can be caused by a decreasing c/a ratio in w-AlN or strains present in the AlN layer. On the other hand, if an unstrained w-AlN layer with the lattice parameters $a = 0.311$ nm and $c = 0.498$ nm [190] is assumed, the smaller difference in ω ($\Delta\omega = 4.4°$) would indicate that the special orientation "w2.1" and "w2.2" as shown in Fig. 7.46b implying twins with the twin boundary $\{101\}_w$ constitute not the prevailing w-AlN columnar grains

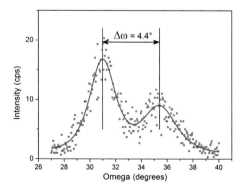

Fig. 7.49: XRD rocking curve of the TiN / AlN / TiN layer measured at $2\theta_B = 66°$ for $\vec{q} \perp$ $[110]_{MgO}$. The blue symbols represent the measured curve and the red line is the fit of two observed maxima. The maxima are separated by approx. 4.4°. The sample position during the measurement of the XRD rocking curve corresponds to Fig. 7.48c.

because then two maxima separated by $\Delta\omega = 6.4°$ would be expected in the rocking curve shown in Fig. 7.49.

However, according to the SAED pattern and the RSM, the special orientations "w2.1" and "w2.2" are among the present columnar grains in the w-AlN layer. For both ORs the $(0\overline{1}\overline{1})_{w2.1}$ and $(0\overline{1}1)_{w2.2}$ planes are parallel to the $(002)_{fcc}$ lattice planes of the TiN seed layer. The lattice misfit along the $[001]_{fcc}$ direction can be calculated from Eq. (7.21) if the interplanar distances of the respective planes of the fcc-phase (d_{fcc}) and the wurtzite phase (d_w) are considered. In case of $d_{fcc}(002)$ and $d_w(0\overline{1}1)$ the misfit is 11 %, if the lattice parameters $a_{fcc} = 0.4241$ nm, $a_w = 0.311$ nm and $c_w = 0.498$ nm [190] are considered.

$$f_{planes} = 2 \cdot \frac{d_{fcc} - d_w}{d_{fcc} + d_w} \qquad (7.21)$$

This relative high misfit of "w2.1" and "w2.2" is probably responsible for the slight rotations of the w-AlN grains around the $[\overline{1}10]_{fcc}$ direction causing further orientations close to "w2.1" and "w2.2" in order to reduce the misfit.

Furthermore, it is assumed that the fcc-TiN seed layer provides a flat interface since no facets were observed using high resolution (HR-) STEM as shown in the HRSTEM DF image in Fig. 7.50a. Thus, the attempt of the w-AlN columnar grains to reduce the misfit along the $[001]_{fcc}$ direction is considered as the main reason for the kind of zig-zag orientation of the adjacent grains leading finally to broadened disc shaped diffraction spots in the measured SAED pattern in Fig. 7.46a.

The possibility of the formation of fcc-AlN (space group $Fm\overline{3}m$) directly at the interface to the TiN seed layer was checked by the analysis of the near edge structure of the nitrogen K edge, because wurtzitic AlN and fcc-AlN with rocksalt structure (space group $Fm\overline{3}m$) can be distinguished by the near edge structure [166, 231]. In order to do this, the EEL spectra around the nitrogen K edge and the titanium $L_{2,3}$ edge were recorded along a line scan with a step size of 1 nm across the interface of the TiN seed layer and the adjacent AlN layer. Simultaneously to the EEL signal, the signal of the energy dispersive X-ray spectroscopy was recorded in order to allocate the position of the interface to the corresponding measuring points. Three selected EEL spectra are shown in Fig. 7.50b which were summated over a measuring distance of 6 nm. The spectrum labelled "TiN" was measured in the TiN seed layer whereas the spectrum labelled "AlN" was measured in the AlN layer. The near edge structure of the nitrogen K edge (see Fig. 7.50c), agrees with the near edge structure for fcc-TiN and wurtzitic AlN obtained from own reference samples (see characteristic features in Fig. 6.7b and Fig. 6.7c or Ref. [166]) or from literature [176, 231]. The spectrum labelled as "TiN / AlN" is the sum of six spectra measured 3 nm before and after the interface. Comparing this spectrum with the fcc-AlN (space group $Fm\bar{3}m$) reference spectrum (see Fig. 6.7a) or the theoretically calculated one [167, 174, 231], no obvious features which are attributed to fcc-AlN can be identified within this spectrum. Instead, typical features of w-AlN, which are labelled as "$A_{w\text{-}AlN}$" and "$B_{w\text{-}AlN}$", are apparent. The same was observed, if each single spectrum close to the interface was checked. Hence, the presence of fcc-AlN ($Fm\bar{3}m$) at the interface was not confirmed.

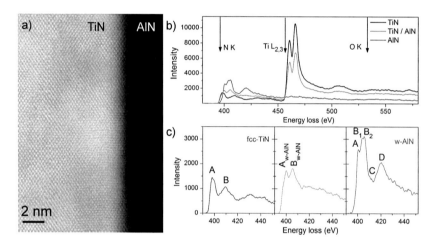

Fig. 7.50: STEM DF image of the sample FIB_2 showing the flat interface between the 1st TiN layer and the adjacent AlN layer (a) and EEL spectra recorded close to the interface in the TiN and AlN layer as well as at the interface showing the N K edge and Ti $L_{2,3}$ edge (b). The N K edge of the spectra is magnified in figure (c) and significant features are marked by A, B, C, D.

In summary, it can be stated that the SAED pattern of sample FIB_2 shown in Fig. 7.46a is caused by w-AlN columnar grains which are oriented close to the orientations "w2.1" and "w2.2" (or "w2.2a") given in the simulation (Fig. 7.46b). Comparing the simulated electron diffraction pattern (Fig. 7.46a) with the measured one (Fig. 7.46b), additional diffraction spots are visible in the SAED pattern. These additional diffraction spots coincidence with the systematically absent reflections due to space group (see open squares in Fig. 7.46b). Due to double diffraction, these diffraction spots could be observed in electron diffraction patterns in TEM [154]. Some of the AlN grains in the AlN layer deposited onto the $(1\bar{1}0)$ oriented fcc-TiN seed layer might be separated by twin boundaries with the $\{101\}_w$ twin boundary being perpendicular to the fcc-TiN / w-AlN interface. Such a twin boundary is shown in Fig. 7.51a. The simultaneous appearance of the orientations "w2.1" and "w2.2" or "w2.2a" in the AlN layer creates a 2 fold rotational symmetry of the w-AlN grains around the $[1\bar{1}0]_{fcc}$ direction of the fcc-TiN seed layer (see Fig. 7.51b). This is obviously necessary to adjust to the TiN seed layer which is characterized by a 2 fold rotation axis in the $[1\bar{1}0]_{fcc}$ direction.

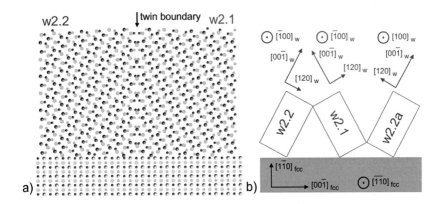

Fig. 7.51: Schematic representation of the w-AlN layer deposited onto the $(1\bar{1}0)$ oriented fcc-MgO substrate as seen in the sample FIB_2. The $[\bar{1}\bar{1}0]_{fcc}$ direction points out of the plane of the paper. Figure (a) shows the atomic arrangement within two w-AlN grains with the orientation "w2.1" and "w2.2" and the twin boundary in between. The Ti atoms are plotted by yellow spheres, the Al atoms by orange spheres and nitrogen by green spheres. Greyish spheres represent atoms which are lying below the plane of the paper. Figure (b) is a sketch of the w-AlN elementary cells with the orientation "w2.1", "w2.2" and "w2.2a" showing their position with respect to the substrate.

The orientations determined in the AlN layer of the sample FIB_2 were checked with the sample **FIB_1** which is related to the sample FIB_2 by a 90° rotation around the $[1\bar{1}0]_{MgO}$ direction (see Fig. 7.42a). SAED done within nearly the whole AlN layer thickness (diameter of the analysed region ~ 97 nm) revealed the electron diffraction pattern given in Fig. 7.52a. The positions of the fcc-TiN seed layer are indicated by the white net in Fig. 7.52a, since the diffraction spots of the fcc-TiN seed layer are not included in the SAED pattern for a better clarity. The theoretical pattern of the fcc-TiN seed layer is labelled by white indices in Fig. 7.52a. The white arrows indicate the orientation of the fcc-MgO substrate which is identical to the orientation of the fcc-TiN seed layer. The SAED pattern of the AlN layer was done with the zone axis $[00\bar{1}]_{fcc}$ in the fcc-TiN seed layer or fcc-MgO substrate.

The SAED pattern of the AlN layer shown in Fig. 7.52a can be indexed with the simulated electron diffraction pattern in Fig. 7.52b where different colours of the simulated diffraction spots mark different orientations. The special orientations "w2.1" ($[\bar{1}\bar{1}0]_{fcc} \parallel [\bar{1}00]_{w2.1}$ and

$(002)_{fcc} \parallel (0\overline{1}\overline{1})_{w2.1}$) and "w2.2" ($[\overline{1}\overline{1}0]_{fcc} \parallel [\overline{1}00]_{w2.2}$ and $(002)_{fcc} \parallel (0\overline{1}1)_{w2.2}$) (or "w2.2a"), which are illustrated in the simulated diffraction pattern of the sample FIB_2, appear in the sample FIB_1 as orientation "w1.1" and "w1.2" (or "w1.2a"), respectively. From the orientations "w1.1" and "w1.2" (or "w1.2a") just the diffraction spots of the $(\overline{2}10)_w$ and $(2\overline{1}0)_w$ planes being parallel to the $(\overline{2}\overline{2}0)_{fcc}$ and $(220)_{fcc}$ planes are visible in the SAED pattern of the sample FIB_1 (see Fig. 7.52a). These diffraction spots are highlighted in the simulated diffraction pattern in Fig. 7.52b by the blue and red marked diffraction spots. The OR of "w1.1", "w1.2" and "w1.2a" to the fcc-TiN seed layer can be described as follows:

"w1.1": $[00\overline{1}]_{fcc} \parallel$ to the normal of the $(011)_{w1.1}$ plane (this corresponds to the direction $[2875, 5750, 1682]_{w1.1}$) and $(\overline{2}\overline{2}0)_{fcc} \parallel (\overline{2}10)_{w1.1}$

"w1.2": $[00\overline{1}]_{fcc} \parallel$ to the normal of the $(01\overline{1})_{w1.2}$ plane (this corresponds to the direction $[2875, 5750, \overline{1682}]_{w1.2}$) and $(\overline{2}\overline{2}0)_{fcc} \parallel (\overline{2}10)_{w1.2}$

"w1.2a": $[001]_{fcc} \parallel$ to the normal of the $(0\overline{1}\overline{1})_{w1.2a}$ plane (this corresponds to the direction $[\overline{2875}, \overline{5750}, \overline{1682}]_{w1.2a}$) and $(\overline{2}\overline{2}0)_{fcc} \parallel (2\overline{1}0)_{w1.2a}$

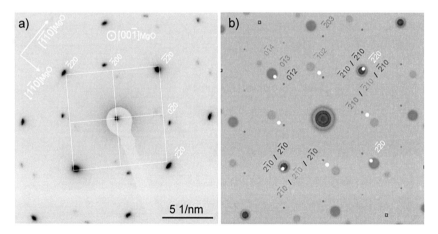

Fig. 7.52: SAED pattern of the AlN layer in the FIB sample 1 (a). The white arrows indicate the orientation with respect to the fcc-MgO substrate. The white lattice marks the diffraction pattern of the fcc-TiN seed layer (not measured) that is labelled by white indices. The measured SAED of the AlN layer can be indexed with the simulated electron diffraction pattern of w-AlN (b) containing the orientations: | 362 |$_w$ (green = w1.3), | 241 |$_w$ (orange = w1.4) and | 121 |$_w$ (grey = w1.5). Additionally, the crystallite groups "w1.1" (red) and "w1.2" (blue) with the $(011)_{w1.1}$ and $(01\overline{1})_{w1.2}$ being parallel to the TEM foil plane contribute to the diffraction pattern (see text for explanation). The white spots indicate the reduced diffraction pattern of fcc-TiN. (b). The open squares correspond to systematically absent reflections due to the space group.

The other diffraction spots visible in the measured SAED pattern are attributed to w-AlN crystallites with the zone axis $[362]_{w1.3}$ (green spots, orientation labelled as "w1.3"), $[241]_{w1.4}$ (orange spots, orientation labelled as "w1.4") and $[121]_{w1.5}$ (grey spots, orientation labelled as "w1.5"). The OR of these three orientations with respect to the fcc-TiN seed layer can be described as follows:

"w1.3": $[00\bar{1}]_{fcc} \parallel [362]_{w1.3}$ and $(\bar{2}\bar{2}0)_{fcc} \parallel (\bar{2}10)_{w1.3}$ and $(2\bar{2}0)_{fcc} \parallel (01\bar{3})_{w1.3}$

"w1.4": $[00\bar{1}]_{fcc} \parallel [241]_{w1.4}$ and $(\bar{2}\bar{2}0)_{fcc} \parallel (\bar{2}10)_{w1.4}$ and $(2\bar{2}0)_{fcc} \parallel (01\bar{4})_{w1.4}$

"w1.5": $[00\bar{1}]_{fcc} \parallel [121]_{w1.5}$ and $(\bar{2}\bar{2}0)_{fcc} \parallel (\bar{2}10)_{w1.5}$ and $(2\bar{2}0)_{fcc} \parallel (01\bar{2})_{w1.5}$

The major orientations in the sample FIB_1 seem to represent the crystallites with the zone axis $[362]_{w1.3}$ (orientation "w1.3") and zone axis $[241]_{w1.4}$ (orientation "w1.4"). Both directions make an angle of 90° with the $[\bar{1}00]_{w2.1/w2.2}$ direction which is the zone axis of sample FIB_2 (see Fig. 7.53). Furthermore, the $[362]_{w1.3}$ direction is parallel to the $(01\bar{3})_{w2.1}$ plane and the $[241]_{w1.4}$ direction intersects the $(01\bar{3})_{w2.1}$ plane at an angle of 6.8° (see Fig. 7.53b). According to the SAED pattern of the sample FIB_2 that is shown in Fig. 7.46a, an orientation distribution of crystallites with the zone axis $[\bar{1}00]_w$ and the $(01\bar{3})_w$ plane being parallel to the interface and inclined up to approx. ± 7° with respect to the interface are present in the sample FIB_2. Thus, the zone axis $[362]_{w1.3}$ and $[241]_{w1.4}$ observed in the sample FIB_1 agree with the orientation distribution observed in the sample FIB_2 with the $[\bar{1}00]_{w2.1}$ zone axis.

The crystallites with the $[121]_{w1.5}$ zone axis (grey spots, orientation labelled "w1.5") seem to represent a minor fraction in the sample FIB_1 because the intensity of the diffraction spots

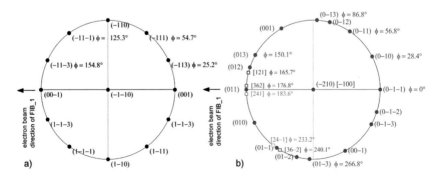

Fig. 7.53: Stereographic projection of the fcc-TiN seed layer (a) and the w-AlN crystallites of the group "w2.1" (b) showing the azimuthal position of selected planes and directions for the polar distance $\psi = 0°$ and $\psi = 90°$. The electron beam direction used for the sample FIB_2 is perpendicular to the plane of the paper. The TiN / AlN interface is parallel to the blue line.

attributed to them is very low as compared to the other orientations with the zone axis $[362]_{w1.3}$ and $[241]_{w1.4}$. The $[121]_{w1.5}$ direction is also perpendicular to the $[\bar{1}00]_{w2.1/2.2}$ direction. According to the observed orientation "w1.5" in the sample FIB_1, the crystallites with the orientation "w1.5" would appear in the sample FIB_2 with the zone axis $[\bar{1}00]_w$ and the $(0\bar{1}3)_w$ plane inclined by 11.1° with respect to the interface. But due to their low volume fraction they are not visible in the SAED pattern of the sample FIB_2 Fig. 7.46a.

This means that the orientations "w1.3", "w1.4" and "w1.5" in the sample "FIB_1" can be deduced from the orientations **close** to "w2.1" in the sample FIB_2. Due to the symmetry within the wurtzite structure, the crystallites **close** to the orientation "w2.2a" and "w2.2" present in the sample FIB_2 would appear in the sample FIB_1 with an identical electron diffraction pattern like "w1.3, "w1.4" and "w1.5". For example the crystallites **close** to the orientation "w2.2a" appear in the sample FIB_1 as "w1.3a, "w1.4a" and "w1.5a" with the following OR to the fcc-TiN seed layer:

"w1.3a": $[00\bar{1}]_{fcc}$ ‖ $[\bar{3}6\bar{2}]_{w1.3a}$ and $(\bar{2}\bar{2}0)_{fcc}$ ‖ $(2\bar{1}0)_{w1.3a}$ and $(2\bar{2}0)_{fcc}$ ‖ $(01\bar{3})_{w1.3a}$

"w1.4a": $[00\bar{1}]_{fcc}$ ‖ $[\bar{2}\bar{4}1]_{w1.4a}$ and $(\bar{2}\bar{2}0)_{fcc}$ ‖ $(2\bar{1}0)_{w1.4a}$ and $(2\bar{2}0)_{fcc}$ ‖ $(01\bar{4})_{w1.4a}$

"w1.5a": $[00\bar{1}]_{fcc}$ ‖ $[\bar{1}2\bar{1}]_{w1.5}$ and $(\bar{2}\bar{2}0)_{fcc}$ ‖ $(2\bar{1}0)_{w1.5a}$ and $(2\bar{2}0)_{fcc}$ ‖ $(01\bar{2})_{w1.5a}$

The identical diffraction pattern as observed by SAED (Fig. 7.52a) was obtained by HRTEM / FFT in local areas in the AlN layer as shown exemplarily in Fig. 7.54 for an area of approx. 23 nm × 23 nm and even in very small areas of approx. 6 nm × 6 nm (not shown). As mentioned above, the width of the AlN columnar grains is about 10 to 20 nm.

The TEM foil thickness was in the case of sample FIB_1 approx. 100 nm as estimated from the EEL spectrum taking into account the total intensity of the spectrum and the intensity of the zero-loss peak according to Eq. (6.63). Thus, it is assumed that the three orientations (w1.3, w1.4 and 1.5) seen in the small areas of the HRTEM image are attributed to the number of columnar grains within the thickness of the TEM foil. Considering the FFT in Fig. 7.54c obtained from an area of 46 nm × 46 nm and the SAED pattern in Fig. 7.52a, a broadened $(0\bar{1}3)_w$ diffraction spot of the crystallites with the $[362]_w$ zone axis is apparent. This indicates a slight rotation of the crystallites around the $[362]_w$ zone axis so that the $(0\bar{1}3)_w$ / $(01\bar{3})_w$ plane of some crystallites is aligned parallel to the interface and of some other crystallites the $(0\bar{1}3)_w$ / $(01\bar{3})_w$ plane is slightly tilted with respect to the interface.

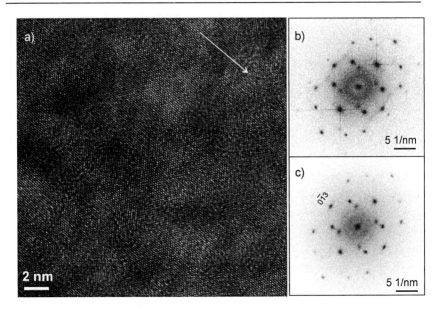

Fig. 7.54: HRTEM image of the AlN layer of the sample FIB_1 (a) and the corresponding FFT of the HRTEM image (b). The FFT of the AlN layer obtained from a four times larger area (c) than the area shown in (a). The white arrow indicates the growth direction being parallel to $[1\bar{1}0]_{MgO}$.

The broadening of the $(01\bar{3})_w$ reciprocal lattice point could be confirmed by RSMs plotted in Fig. 7.55 for $\vec{q}_x \parallel [110]_{MgO}$ and $\vec{q}_z \parallel [1\bar{1}0]_{MgO}$, which were measured at the D8 ADVANCE diffractometer from Bruker AXS. The RSM in Fig. 7.55a shows the broadened reciprocal lattice point attributed to the $(01\bar{3})_w$ plane at $q_x \approx 0$ and $q_z \approx 44.5$ nm^{-1}. The reciprocal lattice point attributed to the $(01\bar{3})_w$ is also visible by a light blue shading at $q_x \approx 0$ and $q_z \approx 44.5$ nm^{-1} in Fig. 7.55b which shows its position in relation to the reciprocal lattice points of $(2\bar{2}0)_{MgO}$ and $(2\bar{2}0)_{TiN}$. The positions of $(2\bar{2}0)_{MgO}$ and $(2\bar{2}0)_{TiN}$ at $q_x \approx 0$ in turn confirm the alignment of the $[1\bar{1}0]_{MgO}$ direction parallel to q_z. Similar to the RSM shown in Fig. 7.48b, an artefact is visible in Fig. 7.55b, which is marked by the white arrow and discussed in the Appendix (see page 197).

The analysis of the XRD rocking curve (see Fig. 7.56) done at $2\theta_B = 66°$ of the 013 diffraction line of w-AlN allowed to describe the slight rotation of the crystallites around the $[362]_w$ direction that was observed by HRTEM / FFT and SAED of the sample FIB_1. The maximum of the rocking curve at $\omega = 32.87°$ and the FWHM of $3.47°$ indicated that the majority of the

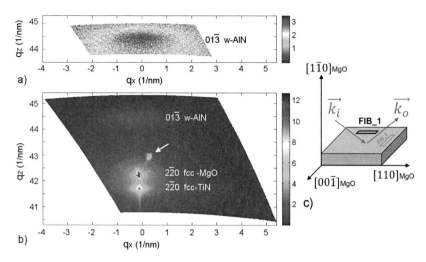

Fig. 7.55: Reciprocal space map of the TiN / AlN / TiN layer stack deposited onto the ($1\bar{1}0$) oriented fcc-MgO single crystal showing the positions and shape of the reciprocal lattice points corresponding to the ($01\bar{3}$)$_w$ plane of w-AlN crystals with the orientation "w1.3" (a) and the ($2\bar{2}0$)$_{fcc}$ plane of the fcc-MgO substrate and the ($2\bar{2}0$)$_{fcc}$ plane of the fcc-TiN seed layer (b). The feature marked by the white arrow in figure (b) indicates an artefact. The colour scale refers to the intensity given as natural logarithm. The orientation of the sample with respect to the incident $\vec{k_i}$ and outgoing $\vec{k_o}$ wave vector during the RSM measurement is displayed in (c) which corresponds to $\vec{q}_z \parallel [1\bar{1}0]_{MgO}$ and $\vec{q}_x \parallel [110]_{MgO}$.

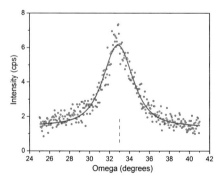

Fig. 7.56: Rocking curve of the TiN / AlN / TiN layer measured at $2\theta_B = 66°$ for $\vec{q} \perp [00\bar{1}]_{MgO}$. The blue symbols represent the measured curve and the red line is the fit of the observed maximum. The grey dashed line indicates the symmetrical position where $\omega = \theta_B$.

crystallites belonging to the $[362]_w$ zone axis of the sample FIB_1 align the $(0\bar{1}3)_w$ / $(01\bar{3})_w$ plane parallel to the interface. This group of crystallites is the one labelled by "w1.3" and indicated by green spots in the simulated electron diffraction pattern in Fig. 7.52b.

All orientations identified in the SAED pattern of the sample FIB_1 ("w1.3", "w1.4" and "w1.5", see Fig. 7.52b) have in common that their $(\bar{2}10)_w$ plane is perpendicular to the interface and parallel to the $(\bar{2}\bar{2}0)_{fcc}$ plane of the TiN seed layer. The misfit of the interplanar spacing along the direction parallel to the interface ($\parallel [\bar{1}\bar{1}0]_{fcc}$) as determined by Eq. (7.21) is just 3.6 %.

According to the SAED pattern observed in the sample FIB_1, the growth of the w-AlN layer on the $(1\bar{1}0)$ oriented fcc-TiN seed layer is summarized in the sketch shown in Fig. 7.57. Due to the low volume fraction of the orientation "w1.5", this orientation is not included in the sketch.

The simultaneous appearance of the orientations "w2.1" and "w2.2" or "w2.2a" as well as "w1.3" and "w1.3a" as well as "w1.4" and "w1.4a" in the AlN layer, creates a 2-fold rotational symmetry of the w-AlN grains around the $[1\bar{1}0]_{fcc}$ direction of the fcc-TiN seed layer (see Fig. 7.51b). In that way the w-AlN layer adjusts to the 2-fold rotation axis along the $[1\bar{1}0]_{fcc}$ direction of the fcc-TiN seed layer.

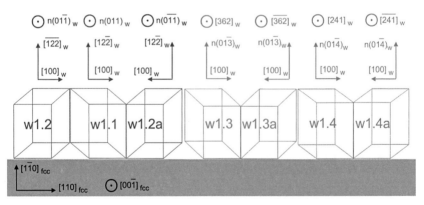

Fig. 7.57: Sketch of the w-AlN elementary cells showing the alignment of the orientation types "w1.1", "w1.2", "w1.2a", "w1.3", "w1.3a", "w1.4" and "w1.4a" with respect to the $(1\bar{1}0)$ oriented fcc-TiN seed layer as seen in the sample FIB_1. The $[00\bar{1}]_{fcc}$ direction points out of the plane of the paper.

Still, the initial question if the OR characterized by $(\bar{1}\bar{1}0)_{fcc} \parallel (\bar{2}10)_w$ and $[001]_{fcc} \parallel [001]_w$ (see Table 7.3 Eq. (7.19)), which was found in the Ti-Al-N monolayer and Ti-Al-N / Al-Ti-(Ru)-N multilayer coatings, is present in the AlN layer grown on the $(1\bar{1}0)_{fcc}$ oriented fcc-TiN seed layer is not answered yet. In case of its presence, the SAED pattern measured in the samples FIB_2 (Fig. 7.46a) and FIB_1 (Fig. 7.52a) should be similar to the simulated electron diffraction patterns shown in the Appendix (see Fig. A. 10). According to the SAED pattern of the AlN layer obtained from the sample FIB_2 (see Fig. 7.46a), the presence of this OR cannot be excluded because the diffraction spots of this OR (see cyan coloured spots in Fig. A. 10a) are overlapping on the one hand with the diffraction spots of the other orientations (w2.1, w2.2) and on the other hand with the spots caused by double diffraction. However, the SAED of the sample FIB_1 did not indicate the presence of the OR since the expected diffraction spots (see cyan coloured spots in Fig. A. 10b) are absent in the measured SAED pattern (see Fig. 7.52a). The absence of the OR given in Eq. (7.19), that was found in the Ti-Al-N monolayers and multilayers, could be attributed to the fact that in the samples FIB_1 and FIB_2 the formation of the above mentioned ORs (w1.1, w1.2, w1.3, w1.4, w1.5, w2.1 and w2.2 (and all "a" versions)) is primarily determined by the orientation of the flat interface of the single crystalline TiN layer and the neighbouring w-AlN columnar grains.

In contrast to the samples FIB_1 and FIB_2, the orientation of the w-AlN crystallites within the CAE Ti-Al-N monolayers and multilayers is determined by several factors. Firstly, the CAE coatings are usually characterized by a higher compressive stress as compared to coatings deposited by magnetron sputtering. Secondly, in Ti-Al-N monolayers and multilayers, the w-AlN crystallites are usually embedded into the fcc-(Ti,Al)N matrix thus forming complex three-dimensional interfaces with the neighbouring fcc phase. Both factors could led to the formation of the OR given in Eq. (7.19) in the CAE Ti-Al-N based layers but not necessarily in the w-AlN layer deposited onto the $(1\bar{1}0)_{fcc}$ oriented fcc-TiN single crystalline seed layer.

However, the heteroepitaxial growth of the w-AlN layer on the $(1\bar{1}0)_{fcc}$ oriented fcc-TiN seed layer showed that the $\{\bar{1}20\}_w$ planes align parallel to the $\{110\}_{fcc}$ planes. The same was found for the OR given in Eq. (7.19). This demonstrates that the alignment of these planes is preferred since the interplanar spacings of the lattice planes $(\bar{1}20)_w$ and $(220)_{fcc}$ match well having just a small misfit of 3.6 %.

7.3.2.2   Wurtzite AlN grown on (00$\overline{1}$)$_{\text{fcc}}$ oriented fcc-TiN

SAED done in the AlN layer of the sample **FIB_3** revealed the electron diffraction pattern given in Fig. 7.58a. As in the case of the previous samples FIB_1 and FIB_2, the diameter of the analysed region was also ~ 97 nm, which covers nearly the whole AlN layer thickness of ~ 112 nm. The grey arrows indicate the orientation of the fcc-MgO substrate being identical to the orientation of the fcc-TiN seed layer. The SAED pattern of the AlN layer was measured while maintaining the zone axis [$\overline{1}\overline{1}0$]$_{\text{fcc}}$ of the fcc-TiN seed layer or fcc-MgO substrate which were not included in the selected area of the diffraction experiment. However, the theoretical positions of the diffraction spots of the fcc-TiN seed layer are indicated by the white lattice in Fig. 7.58a and are labelled by grey indices. The interface between the fcc-TiN seed layer and the AlN layer is parallel to the [$1\overline{1}0$]$_{\text{MgO}}$ direction and the growth direction of the layer stack is parallel to the [$00\overline{1}$]$_{\text{MgO}}$ direction.

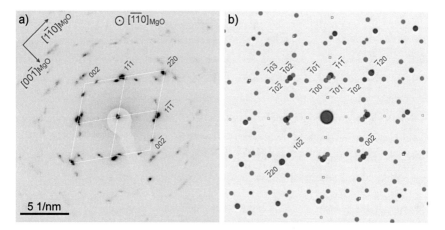

Fig. 7.58: SAED pattern of the AlN layer in the FIB sample 3 (a). The grey arrows indicate the orientation with respect to the fcc-MgO substrate. The white lattice marks the diffraction pattern of the fcc-TiN seed layer that is labelled by grey indices. Simulated electron diffraction pattern of w-AlN (b) showing three orientations (red = w3.1, blue = w3.2 and green = w3.3). The grey spots indicate the reduced diffraction pattern of fcc-TiN (b). The open squares correspond to systematically absent reflections in wurtzite due to the space group.

The SAED pattern of the w-AlN layer can be described by three main orientations that are characterized by a slight misorientation leading to the disc shape broadening of the diffraction spots (see Fig. 7.58a) similar to the SAED patterns of the samples FIB_1 and FIB_2. The three main orientations can be indexed with the simulated electron diffraction pattern in Fig. 7.58b where different colours of the diffraction spots indicate different orientations of the w-AlN crystallites.

In Fig. 7.58b the diffraction spots marked by red colour represent the w-AlN crystallites with the orientation "w3.1" with the zone axis $[\bar{2}\bar{1}1]_{w3.1}$. The diffraction spots marked by the blue and green colour can be attributed to the w-AlN crystallites with the orientation "w3.2" and "w3.3" and the zone axis $[0\bar{1}0]_{w3.2}$ and $[010]_{w3.3}$, respectively. The OR of these three orientations with respect to the fcc-TiN seed layer can be described as follows:

"w3.1": $[\bar{1}\bar{1}0]_{fcc} \parallel [\bar{2}\bar{1}1]_{w3.1}$ and $(2\bar{2}0)_{fcc} \parallel (\bar{1}20)_{w3.1}$ and $(00\bar{2})_{fcc} \parallel (102)_{w3.1}$

"w3.2": $[\bar{1}\bar{1}0]_{fcc} \parallel [0\bar{1}0]_{w3.2}$ and $(2\bar{2}0)_{fcc} \parallel (\bar{1}02)_{w3.2}$

"w3.3": $[\bar{1}\bar{1}0]_{fcc} \parallel [010]_{w3.3}$ and $(2\bar{2}0)_{fcc} \parallel (10\bar{2})_{w3.3}$

The crystallites with the orientation "w3.2" can be transformed to the crystallites "w3.3" by a 180° rotation around the $[00\bar{1}]_{fcc}$ direction which is parallel to the $[00\bar{1}]_{MgO}$ direction.

The green marked diffraction spots in the simulated diffraction pattern in Fig. 7.58b can be also obtained from the orientation "w3.2" (blue spots) by a counter-clockwise rotation of 85.5° around the $[0\bar{1}0]_{w3.2}$ zone axis leading to the orientation "w3.3a". After this rotation the $(\bar{1}0\bar{2})_{w3.3a}$ plane is on the position of the $(10\bar{2})_{w3.2}$ plane in Fig. 7.58b. The OR labelled as "w3.3a" can be described as:

"w3.3a": $[\bar{1}\bar{1}0]_{fcc} \parallel [0\bar{1}0]_{w3.3a}$ and $(2\bar{2}0)_{fcc} \parallel (102)_{w3.3a}$

Considering the theoretical calculations by Béré *et al.* [228, 229], the $\{102\}_w$ plane was identified as a possible twin boundary in wurtzite structures. Furthermore, Gindt and Kern [232] could observe the $(10\bar{2})_w$ twin boundary in an industrially produced w-AlN crystal that was grown at high temperatures. Hence, it is deduced that twins with the $\{102\}_w$ twin boundary can be formed within the AlN layer of the sample FIB_3. For the orientations "w3.2" and "w3.3" or "w3.2" and "w3.3a" plotted in the simulated electron diffraction pattern in Fig. 7.58b, the twin boundary would be perpendicular to the interface and would correspond e.g. to the $(\bar{1}0\bar{2})_{w3.2}$ plane in Fig. 7.58b. In that case, the lattice planes $(\bar{1}0\bar{2})_{w3.2}$ and $(\bar{1}02)_{w3.3a}$ would embed an angle of 9° whereby the bisecting line is parallel to the $[00\bar{1}]_{MgO}$ direction which is parallel to the normal of the interface.

All orientations described above (w3.1, w3.2 and w3.3 / w3.3a) have in common that the $\{102\}_w$ plane is nearly parallel to the interface. This could be supported by the RSM of the TiN / AlN / TiN layer stack that was measured in the vicinity of the 102 diffraction line and plotted in Fig. 7.59a for $\vec{q}_x \parallel [1\bar{1}0]_{MgO}$ and $\vec{q}_z \parallel [00\bar{1}]_{MgO}$ in the q_x-range of -5 nm$^{-1} \leq q_x \leq 5$ nm^{-1}. This RSM revealed three broadened reciprocal lattice points attributed to the

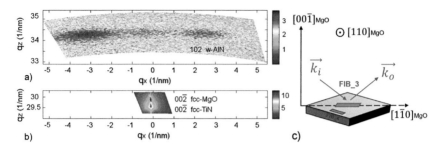

a)

b)

c)

Fig. 7.59: Reciprocal space map of the TiN / AlN / TiN layer stack deposited onto the $(00\bar{1})$ oriented fcc-MgO single crystal showing the positions and shape of the reciprocal lattice points corresponding to $\{102\}_w$ planes of the w-AlN crystals (a) and the $(00\bar{2})_{fcc}$ plane of the fcc-MgO substrate and the $(00\bar{2})_{fcc}$ plane of the fcc-TiN seed layer (b). The colour scale refers to the intensity given as natural logarithm. The orientation of the sample with respect to the incident $\vec{k}_i$ and outgoing $\vec{k}_o$ wave vector during the RSM measurement is displayed in (c) which corresponds to $\vec{q}_z \parallel [00\bar{1}]_{MgO}$ and $\vec{q}_x \parallel [1\bar{1}0]_{MgO}$.

$\{102\}_w$ plane. The three broadened reciprocal space points could be assigned to three groups of crystallites with the orientation close to "w3.1", "w3.2" and "w3.3" / "w3.3a". Each group of crystallites is characterized by a slight misorientation around the $[\bar{1}\bar{1}0]_{MgO}$ direction which leads to the broadening of the reciprocal lattice points in the measured RSM (Fig. 7.59a) and the broadened diffraction spots in the SAED pattern (Fig. 7.58a).

The three groups of crystallites are better visible in the rocking curve shown in Fig. 7.60 that was measured at $2\theta_B = 49.82°$ showing three maxima. The second maxima in the rocking curve

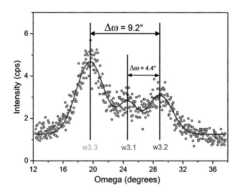

Fig. 7.60: Rocking curve of the TiN / AlN / TiN layer measured at $2\theta_B = 49.82°$. The blue symbols represent the measured curve and the red curve is the fit of three observed maxima. The black arrows indicate the difference in ω of the observed maxima. The maxima could be attributed to the orientations "w3.1", "w3.2" and "w3.3".

(Fig. 7.60) can be assigned to the crystallites with the "w3.1" orientation with the $\{102\}_w$ plane parallel to the interface.

The first and third maxima are separated by $\Delta\omega = 9.2°$ and represent $\{102\}_w$ lattice planes that are tilted by approx. $\pm 4.5°$ from the sample surface. Thus, they could be attributed to the crystallite groups "w3.2" and "w3.3" / "w3.3a".

The misfit along the $[1\bar{1}0]_{fcc}$ direction as calculated by Eq. (7.21) and taking into account the interplanar distances between the $(2\bar{2}0)_{fcc}$ lattice planes of the fcc-TiN seed layer and the lattice planes of the adjacent w-AlN columnar grains is just 3.6 % for the orientation "w3.1" with the lattice planes $(\bar{1}20)_{w3.1}$ being perpendicular to the substrate surface. In contrast to that, the orientation "w3.2" / "w3.3" are characterized by a huge misfit of 19.8 % between the lattice planes $\{\bar{1}02\}_w$ and $(2\bar{2}0)_{fcc}$ along the $[1\bar{1}0]_{fcc}$ direction.

But if the misfit of these crystallites ("w3.2" / "w3.3") along the $[\bar{1}\bar{1}0]_{fcc}$ direction, which is a direction also parallel to the fcc-TiN / w-AlN interface but 90° rotated with respect to the $[1\bar{1}0]_{fcc}$ direction (see Fig. 7.71 at the end of this Chapter), is considered, the misfit is just 3.6 %. This misfit is determined by the lattice planes $(\bar{1}20)_{w3.2}$ and $(\bar{2}\bar{2}0)_{fcc}$ which are perpendicular to the interface along the $[\bar{1}\bar{1}0]_{fcc}$ direction.

The basic orientations assigned in the SAED pattern of the w-AlN layer of the sample FIB_3 (see Fig. 7.58) could be confirmed by HRTEM / FFT (see Fig. 7.61). The columnar grain structure that was apparent in the TEM BF image (Fig. 7.45a) is also visible in the HRTEM image. The width of the columns is approx. 10 nm. The thickness of the TEM foil is approx. 40 nm as estimated from the EELS zero loss signal using Eq. (6.63).

Thus, several columns are expected within the TEM foil thickness. Nevertheless, local FFT done in an apparent columnar grain of the w-AlN layer and marked by the green square (size: 11 nm × 11 nm) in Fig. 7.61a revealed w-AlN crystals close to the orientations "w3.1" and "w3.2" (Fig. 7.61b,c). Local FFT done in the neighbouring w-AlN column and marked by the blue square could be assigned to the crystals close to the orientations "w3.1" and "w3.3" (Fig. 7.61d,e). The FFT from the area marked by the orange square and with the size of 23 nm × 23 nm which contains the two adjacent columnar grains equals the diffraction pattern obtained by SAED (compare Fig. 7.58a and Fig. 7.61f). The broadening of the diffraction spots in the FFT in Fig. 7.61f (e.g. $(\bar{1}20)_{w3.1}$ or $(\bar{1}03)_{w3.2}$) indicates a slight rotation of the three basic orientations around the $[\bar{1}\bar{1}0]_{fcc}$ direction. Thus, it is deduced that at least the three different orientations of the w-AlN columnar grains and their slight rotation are obviously necessary to reduce the misfit between the w-AlN layer and the fcc-TiN seed layer.

As in the case of the TiN / AlN / TiN layer deposited onto $(1\bar{1}0)_{fcc}$ oriented fcc-MgO single crystal, the fcc-TiN seed layer provided a flat interface as shown in the HRSTEM BF image in Fig. 7.62a since no facets were observed. On the basis of ELNES of the N K edge that was measured along a line scan across the TiN / AlN interface with a step size of 1 nm, the formation of fcc-AlN (space group Fm$\bar{3}$m) at the interface could not be confirmed. This is illustrated by

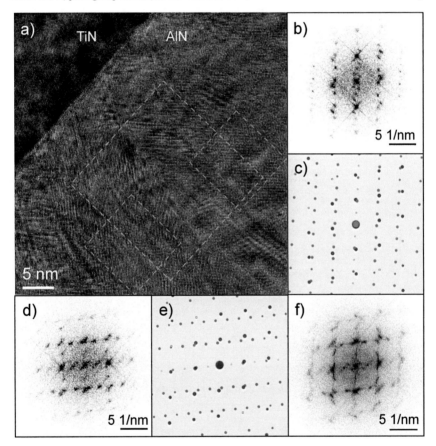

Fig. 7.61: HRTEM image of the TiN / AlN interface in the sample FIB_3 (a). The FFT in (b) originates from the green marked square and corresponds to the simulated diffraction pattern in (c). The FFT in (d) originates from the blue marked square and corresponds to the simulated diffraction pattern in (e). The FFT in (f) was obtained from the orange marked square. The different colours of the spots in the simulated diffraction pattern indicate different orientations of w-AlN: $[1\bar{1}\bar{1}]_w$ (red), $[010]_w$ (green) and $[0\bar{1}0]_w$ (blue). The FFTs and the simulations in the subfigures were rotated in that manner that the direction of the interface is aligned parallel to the interface visible in the HRTEM image.

the EEL spectra in Fig. 7.62b,c. The position of the interface was allocated from the signal of the energy dispersive X-ray spectroscopy, which was recorded simultaneously to the signal of EELS.

The spectrum labelled "TiN" is the sum of six adjacent spectra in the TiN seed layer. The shape of the N K edge indicated fcc-TiN. The spectrum labelled "AlN" is the sum of six adjacent spectra that were measured in the AlN layer. It confirms the wurtzitic phase of w-AlN since the spectrum contains the features typical for w-AlN (see features "A", "B_1", "B_2" in Fig. 6.7b). The spectrum labelled "TiN / AlN" is the sum of six adjacent spectra that were measured across the interface. The specific features of fcc-AlN could not be recognized in the spectrum. Instead typical features of the wurtzite phase, like the features "A", "B_1", "B_2" from Fig. 6.7b are obvious.

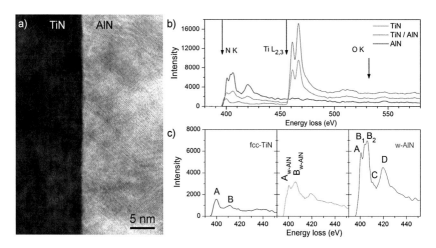

Fig. 7.62: STEM BF image of the sample FIB_3 showing the flat interface between the fcc-TiN seed layer and the adjacent AlN layer (a) and EEL spectra recorded close to the TiN and AlN layer as well as at the TiN / AlN interface showing the N K edge and the Ti $L_{2,3}$ edge (b). The N K edge of each spectrum is magnified in figure (c) and significant features are marked by A, B, C, D.

The comparison of the SAED pattern obtained in the sample FIB_3 with the SAED pattern measured in the sample FIB_2 showed the influence of the orientation of the fcc-TiN seed layer. In both samples the zone axis of the fcc-TiN seed layer was $[\overline{1}\overline{1}0]_{fcc}$ but in the sample FIB_2 the TiN / AlN interface is perpendicular to the $[1\overline{1}0]_{fcc}$ direction and in the sample FIB_3 the TiN / AlN interface is perpendicular to the $[00\overline{1}]_{fcc}$ direction. The diffraction pattern of both samples were different (cf. Fig. 7.46a and Fig. 7.58a) which emphasizes that the formed orientation relationships between the fcc phase and the wurtzite phase are determined by the orientation of the fcc-TiN seed layer. In case of the sample FIB_2, the fcc-TiN seed layer is characterized by a 2 fold rotation axis being normal to the interface, whereas in the sample FIB_3, the fcc-TiN seed layer is characterized by a 4 fold rotation axis being normal to the interface.

The orientations determined in the AlN layer of the sample FIB_3 were checked with the sample **FIB_4** which is related to the sample FIB_3 by a 45° rotation around the $[00\overline{1}]_{MgO}$ direction (see Fig. 7.42b). SAED done in the AlN layer of the sample **FIB_4** revealed the electron diffraction pattern given in Fig. 7.63a. Like in the previous FIB samples the diameter of the analysed region was also ~ 97 nm which covers nearly the whole AlN layer thickness of ~ 112 nm. The grey arrows indicate the orientation of the fcc-MgO substrate. Its orientation is equal to the fcc-TiN seed layer. The diffraction spots of the fcc-TiN seed layer are not included in the SAED pattern, but the white lattice in Fig. 7.63a indicates their theoretical positions which are labelled by grey indices. SAED done in the AlN layer was done with the electron beam direction being parallel to the $[0\overline{1}0]_{MgO}$ direction. The growth direction of the layer stack is parallel to the $[00\overline{1}]_{MgO}$ direction and the interface between the fcc-TiN seed layer and the AlN layer is parallel to the $[100]_{MgO}$ direction.

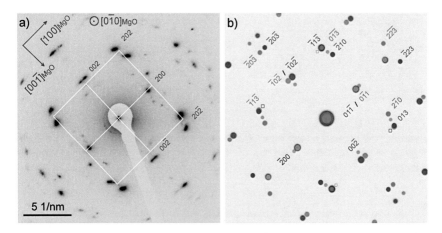

Fig. 7.63: SAED pattern of the AlN layer in the sample FIB_4 (a). The grey arrows indicate the orientation with respect to the fcc-MgO substrate. The white lattice marks the diffraction pattern of the fcc-TiN seed layer with the corresponding indices plotted in grey. Simulated electron diffraction pattern of w-AlN (b) showing four orientations (red = w4.1, blue = w4.2, green = w4.3 and orange = w4.4). The grey spots indicated the reduced diffraction pattern of fcc-TiN (b). The open squares correspond to systematically absent reflections in wurtzite due to the space group.

Several orientations are present in the SAED pattern. In the simulated electron diffraction pattern in Fig. 7.63b four main orientations are displayed which could be identified in the SAED pattern in Fig. 7.63a. Each colour in the simulated diffraction pattern represents a certain orientation of the w-AlN crystallites: red = w4.1, blue = w4.2, green = w4.3 and orange = w4.4 that can be described as follows:

"w4.1": $[0\bar{1}0]_{fcc}$ ‖ $[\bar{2}11]_{w4.1}$ and $(200)_{fcc}$ ‖ $(01\bar{1})_{w4.1}$

"w4.2": $[0\bar{1}0]_{fcc}$ ‖ $[\bar{3}62]_{w4.2}$ and $(004)_{fcc}$ nearly ‖ $(\bar{2}0\bar{3})_{w4.2}$

"w4.3": $[0\bar{1}0]_{fcc}$ ‖ $[36\bar{2}]_{w4.3}$ and $(004)_{fcc}$ nearly ‖ $(\bar{2}0\bar{3})_{w4.3}$

"w4.4": $[0\bar{1}0]_{fcc}$ ‖ $[2\bar{1}\bar{1}]_{w4.4}$ and $(200)_{fcc}$ ‖ $(0\bar{1}1)_{w4.4}$

Taking into account the cutting angle of approx. 45° between the sample FIB_3 and FIB_4 (see Fig. 7.42b), the orientations "w3.1", "w3.2" and "w3.3" that were observed in the sample FIB_3 lead to the orientations "w4.1", "w4.2" and "w4.3" in the sample FIB_4 since the angles between the electron beam directions in both samples are close to 45°, namely: $\angle([\bar{2}11]_{w3.1}$, $[\bar{2}11]_{w4.1}) = 40°$, $\angle([0\bar{1}0]_{w3.2}, [\bar{3}62]_{w4.2}) = 42.5°$ and $\angle([010]_{w3.3}, [36\bar{2}]_{w4.3}) = 42.5°$.

The orientation "w4.4" that is visible in the SAED pattern of the sample FIB_4 can be deduced from the orientation "w3.4" with the zone axis $[21\bar{1}]_{w3.4}$ in the FIB sample 3. The angle between the zone axis of both orientations ($[21\bar{1}]_{w3.4}$, $[2\bar{1}\bar{1}]_{w4.4}$) corresponds to 40° whereby the orientation "w3.4" can be described by:

"w3.4": $[\overline{110}]_{fcc}$ ‖ $[21\bar{1}]_{w3.4}$ and $(2\bar{2}0)_{fcc}$ ‖ $(1\bar{2}0)_{w3.4}$

The orientation "w3.4" can be obtained by a 180° rotation of the orientation "w3.1" around the surface normal leading to the identical diffraction spots like the orientation "w3.1" in the SAED pattern of the sample FIB_3 (see red marked spots in Fig. 7.58b). Thus, it was not identified in the SAED pattern of the sample FIB_3 (see Fig. 7.58) before analysing the SAED pattern of the sample FIB_4.

Furthermore, the orientation "w4.1" can be transformed to the orientation "w4.4" by an 180° rotation around the normal of the fcc-TiN / w-AlN interface which agrees with the $[00\bar{1}]_{MgO}$ direction. In that way, the plane $(01\bar{1})_{w4.1}$ takes the position of the plane $(0\bar{1}1)_{w4.4}$. As mentioned above in Section 7.3.2.1, Béré et al. [228, 229] proposed by theoretical calculations the {101} planes beside the {102} and {103} planes as possible twin boundaries in wurtzite structures and Horiuchi et al. [230] observed the {101} plane as a twin boundary in w-AlN whiskers. Hence, it is deduced that twins can be formed in the w-AlN layer leading to the orientations "w4.1" and "w4.4". In the simulated electron diffraction pattern shown in Fig. 7.63b the twin boundary would correspond for instance to the $(01\bar{1})_{w4.1}$ plane. Thus, the twin boundaries would be perpendicular to the interface and could constitute the column boundaries.

FFT / HRTEM of a local area (see Fig. 7.64) with a size of 23 nm × 23 nm confirmed also the orientations "w4.1", "w4.2", "w4.3" and "w4.4" that were found in the SAED pattern in Fig. 7.63a. Furthermore, additional orientations became apparent in the FFT / HRTEM (see highlighted spots in Fig. 7.64b and FFTs of Fig. 7.66) which are just weakly visible in the SAED pattern in Fig. 7.63a. These orientations ("w4.5" to "w4.8") are plotted additionally to the orientations "w4.1" to "w4.4" by cyan ("w4.5"), lime ("w4.6"), black ("w4.7") and grey ("w4.8") colour in Fig. 7.65. The zone axes are $[\bar{2}31]_{w4.5}$, $[23\bar{1}]_{w4.6}$, $[4\bar{3}2]_{w4.7}$ and $[\bar{4}3\bar{2}]_{w4.8}$. These four zone axes are parallel to the $[0\bar{1}0]_{fcc}$ direction of the fcc-TiN seed layer.

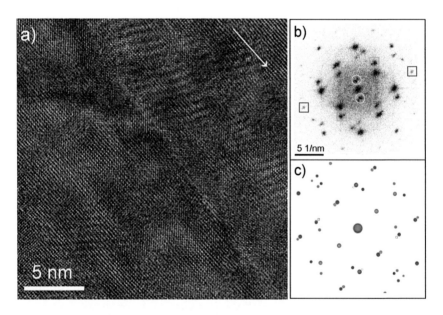

Fig. 7.64: HRTEM image of the w-AlN layer in the sample FIB_4 (a) and corresponding FFT (b). The FFT agrees with the simulated diffraction pattern in (c) (for indices see Fig. 7.63b). Additional reflections in the FFT correspond to reflections caused by Moiré fringes (yellow circles) and further orientations $[\bar{2}31]_{w4.5}$ (cyan squares) and $[4\bar{3}2]_{w4.7}$ (black squares). The white arrow in figure (a) indicates the growth direction of the film.

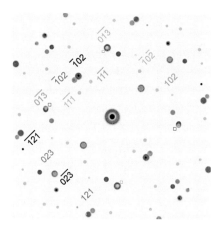

Fig. 7.65: Simulated electron diffraction pattern of the w-AlN layer of the sample FIB_4 showing eight orientations: red = w4.1, blue = w4.2, green = w4.3, orange = w4.4, cyan = w4.5, lime = w4.6, black = w4.7 and grey = w4.8. The indices refer to the orientations w4.5 to w4.8.

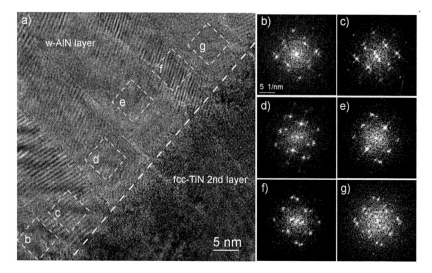

Fig. 7.66: HRTEM image of the w-AlN layer in the sample FIB_4 showing different w-AlN columns (a) and the corresponding FFTs (b-g). The FFTs in the subfigures (b-g) were rotated in that manner that the direction of the interface is aligned parallel to the w-AlN / fcc-TiN interface visible in the HRTEM image.

The different orientations can be assigned to w-AlN columns with a width of approx. 10 nm. This can be deduced from the analysis of the HRTEM image in Fig. 7.66. However, since the thickness of the TEM foil is approx. 60 nm, as estimated from the EEL spectrum according to Eq. (6.63), it is assumed that several w-AlN columns are super-imposed across the thickness of the TEM foil. Thus, several orientations appeared in the individual FFTs that were obtained from an individual area with a size of 6 nm × 6 nm as indicated by the dashed squares in Fig. 7.66a. The analysis of the FFTs showed that the marked regions (b to f) in Fig. 7.66a contain w-AlN columns with orientations close to: "w4.1" + "w4.3" + "w4.5" (region b); "w4.3" + "w4.4" + "w4.5" (region c); "w4.2" + "w4.4" + "w4.6" (region d); "w4.1" + "w4.2" + "w4.6" (region e); "w4.2" + "w4.4" + "w4.6" (region f) and "w4.1" + "w4.2" + "w4.3" + "w4.5" + "w4.6" (region g).

The presence of the orientations "w4.7" and "w4.8" that were visible in the SAED pattern of the sample FIB_4 (Fig. 7.63a, Fig. 7.65) cannot be excluded in the regions b to g in the HRTEM micrograph shown in Fig. 7.66. However, the diffraction spots $(\overline{1}2\overline{1})_{w4.7}$ and $(121)_{w4.8}$ that are characteristic for the orientations "w4.7" and "w4.8" (see Fig. 7.65) do not provide sufficient intensity in the FFTs that are obtained from such small areas like in Fig. 7.66a.

The above mentioned orientations ("w4.1" and "w4.8"), which are considered to be necessary for the reduction of the misfit to the bottom layer, could be identified in the AlN layer by SAED (Fig. 7.63) and HRTEM / FFT (Fig. 7.64, Fig. 7.66). In case of the $(00\overline{1})$ oriented fcc-TiN seed layer obviously more orientations of w-AlN are necessary in order to reduce the misfit than in the case of the $(1\overline{1}0)$ oriented fcc-TiN seed layer.

However, the inspection of the interface region between the $(00\overline{1})$ oriented fcc-TiN seed layer and the adjacent w-AlN layer using HRTEM / FFT, as shown in Fig. 7.67, indicated a transition zone with a width of maximum 7 nm. The FFTs done in the TiN layer (Fig. 7.67b) and over the TiN / AlN interface (Fig. 7.67c) covering an area of 11 nm × 11 nm showed quite similar diffraction patterns. The comparison of the FFTs that were obtained from a sample area with a size of 6 nm × 6 nm in the TiN layer (Fig. 7.67d) and in the AlN layer (Fig. 7.67e) closely to the interface showed also similar diffraction patterns but with blurred diffraction spots in the AlN layer.

The diffraction pattern of TiN agrees with the $[0\overline{1}0]_{fcc}$ zone axis and can be indexed with the indices shown in Fig. 7.67b. However, the indexing of the pattern shown in Fig. 7.67e, which was obtained in the AlN layer, is ambiguous.

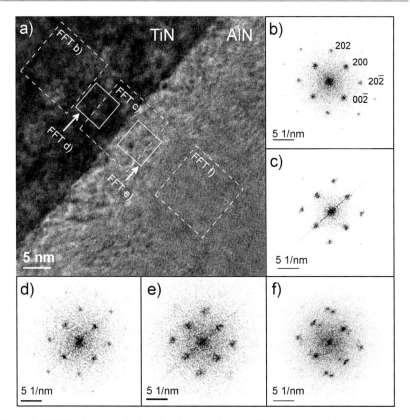

Fig. 7.67: HRTEM image of the interface between the fcc-TiN seed layer and the adjacent AlN layer in the sample FIB_4 (a) with the corresponding FFTs (b-f) obtained in different distances from the interface. The FFTs (b to f) were rotated in that manner that the direction of the interface is aligned parallel to the interface visible in the HRTEM image.

The diffraction pattern could be assigned possibly to fcc-AlN with the same orientation as the fcc-TiN seed layer or to a distorted w-AlN whereby the zone axis $[23\bar{1}]_w$ or $[211]_w$ would be possible. In order to distinguish between w-AlN and fcc-AlN, EELS of the N K edge was done over the TiN / AlN interface. The near edge structure of the N-K edge in the EEL spectra that were obtained from very local areas close to the interface in the sample FIB_4 resembled to the one typical for w-AlN [167, 176, 231] (cf. Fig. 6.7b). This was also confirmed by EELS done in the sample FIB_3 (see Fig. 7.62) that was obtained from the same TiN / AlN / TiN layer stack deposited onto $(00\bar{1})$ oriented MgO. Thus, the formation of fcc-AlN directly at the TiN / AlN interface is excluded and the formation of a transition zone consisting of distorted w-AlN is considered to be most likely. Behind this transition zone the w-AlN layer grows with the orientations close to "w4.1", "w4.2", "w4.3", "w4.4", "w4.5", "w4.6", "w4.7" and "w4.8" as described above. For instance in the selected area of the HRTEM image in Fig. 7.67 the orientations close to "w4.1, "w4.2", "w4.4" and "w4.6" could be deduced from the FFT in Fig. 7.67f.

As in the case of the previous samples FIB_1, FIB_2 and FIB_3, the diffraction pattern obtained by SAED and FFT / HRTEM from the w-AlN layer of the sample FIB_4 is characterized by broadened diffraction spots (see Fig. 7.63a and Fig. 7.64b). This indicates a slight rotation of the w-AlN crystallites around the $[0\bar{1}0]_{MgO}$ direction which is obviously also necessary to reduce the misfit between the w-AlN columns and the bottom layer.

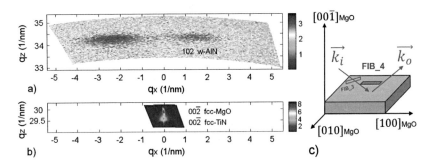

Fig. 7.68: Reciprocal space map of the TiN / AlN / TiN layer stack deposited onto the $(00\bar{1})$ oriented fcc-MgO single crystal showing the positions and shape of the reciprocal lattice points corresponding to $\{102\}_w$ planes of the w-AlN crystals (a) and the $(00\bar{2})_{fcc}$ plane of the fcc-MgO substrate and the $(00\bar{2})_{fcc}$ plane of the fcc-TiN seed layer (b). The colour scale refers to the intensity given as natural logarithm. The orientation of the sample with respect to the incident $\vec{k}_i$ and outgoing $\vec{k}_o$ wave vector during the RSM measurement is displayed in (c) which corresponds to $\vec{q}_z \parallel [00\bar{1}]_{MgO}$ and $\vec{q}_x \parallel [100]_{MgO}$.

The RSM shown in Fig. 7.68a with $\vec{q}_x \parallel [100]_{MgO}$ and $\vec{q}_z \parallel [00\bar{1}]_{MgO}$ could confirm this broadening for the 102 reciprocal lattice points originating from the orientations "w4.1", "w4.4", "w4.7" and "w4.8" (cf. Fig. 7.63b and Fig. 7.65). Comparing the RSM measured for the $00\bar{2}$ reciprocal lattice points of the fcc-MgO substrate and the fcc-TiN seed layer (Fig. 7.68b) with the RSM of the w-AlN layer (Fig. 7.68a), it is visible that the majority of the $\{102\}_w$ planes is inclined with respect to the sample surface. This is better visible in the rocking curve measured at $2\theta_B = 49.82°$ for the 102 diffraction line of w-AlN (see Fig. 7.69) which revealed two maxima that are separated by 6.3°. The first maximum can be attributed to the crystallites close to the orientations "w4.4" and "w4.8". The second diffraction maxima corresponds to crystallites close to the orientation "w4.1" and "w4.7". Both are inclined by approx. 3° from the $[00\bar{1}]_{fcc}$ direction.

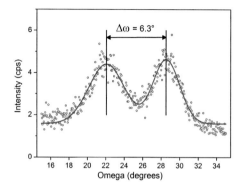

Fig. 7.69: Rocking curve of the TiN / AlN / TiN layer measured at $2\theta_B = 49.82°$. The blue symbols represent the measured curve and the curve is the fit of two observed maxima. The black arrow indicates the difference in ω of the observed maxima.

On the basis of the above presented results, the w-AlN layer deposited onto the $(00\bar{1})_{fcc}$ oriented fcc-TiN seed layer can be described by the following simplified model shown in Fig. 7.70. This model illustrates the arrangement of the individual w-AlN columns that are present across the TEM foil thickness of the sample FIB_4 and that are shown in the HRTEM micrograph in Fig. 7.66. In each area labelled as b, c, d, e, f and g in the HRTEM image in Fig. 7.66a three to five different orientations could be assigned by local FFT. Thus, at least three w-AlN columns are superimposed in each area. Since the smallest apparent column width visible in the HRTEM image (Fig. 7.66) was approx. ~ 10 nm and the thickness in the TEM foil is ~ 60 nm, up to six overlapping columns could be possible in each area ("b" to "g"). The fourth, fifth and sixth column could have the same orientation as the previous three columns or could have the orientation "w4.7" or "w4.8" which cannot be identified from the FFTs of areas as small as 6 nm × 6 nm.

In Fig. 7.70 a possible arrangement of the columns with the orientations close to the orientations "w4.1" to "w4.6" is given. The zone axes of these orientations are parallel to the $[0\bar{1}0]_{MgO}$ direction. Since the orientations of the w-AlN layer are characterized by a slight rotation around the $[0\bar{1}0]_{MgO}$ direction, as indicated by the broadened diffraction spots in the SAED pattern (Fig. 7.63a) and RSM (Fig. 7.68), the labels "w4.1" to "w4.6" in Fig. 7.70 refer also to orientations that are close to the basic orientations "w4.1" to "w4.6" (as given on page 157). As a consequence of this, for instance the neighbouring columns labelled as "b" and "c" with the orientation label "w4.4" in Fig. 7.70 might be separated by a low angle boundary or could have the identical orientation. The low resolution of the FFTs in Fig. 7.66 did not permit a reliable determination of small misorientations. Thus, possible low angle boundaries are indicated by dashed lines in Fig. 7.70. An additional sketch showing the alignment of the w-AlN crystals with respect to the $(0\overline{1}0)$ oriented fcc-TiN seed layer, as seen in the SAED pattern of the sample FIB_4, is given in the Appendix in Fig. A. 11

According to the model, the columns whose basic orientation differ by a rotation of 180° around the growth direction of the film ($[00\overline{1}]_{MgO}$) e.g. "w4.1" and "w4.4" or "w4.2" and "w4.3" or

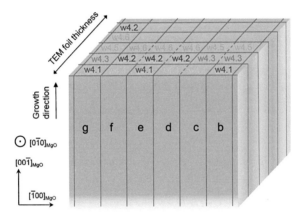

Fig. 7.70: Simplified modell of the w-AlN layer grown on $(00\overline{1})_{fcc}$ oriented fcc-TiN seed layer which is shown in the HRTEM micrograph in Fig. 7.66a. The letters b to g refer to the columns seen in the HRTEM image. The orientations of individual columns are labelled by "w4.1" to "w4.6". Their zone axis is parallel to the view direction $[0\overline{1}0]_{MgO}$.

"w4.5" and "w4.6" are neighbouring columns. Such an arrangement means in case of the w-AlN columns with the orientations close to "w4.1" and "w4.4" that twin boundaries with the twin plane $\{101\}_w$ [228-230] can form as column boundaries. In case of the w-AlN columns with the orientations close to the orientations "w4.2" and "w4.3" that were observed in the sample FIB_3 as the orientations "w3.2" and "w3.3", twin boundaries with the twin plane $\{102\}_w$ [228, 229, 232] can form as column boundaries. Such kind of column boundaries contributes to a lower energy of the boundaries.

However, a closer inspection of the arrangement of the columns within the w-AlN layer would be possible from a TEM foil prepared in plane view orientation with the plane of the foil being parallel to the substrate surface, which corresponds to the $(00\bar{1})_{MgO}$ plane according to the definition used within this Section 7.3 (see Fig. 7.42b).

Finally, the observed orientations of the w-AlN columnar grains in the w-AlN layer deposited onto the $(00\bar{1})$ oriented fcc-TiN seed layer can be explained by the symmetry of the fcc-TiN seed layer. The fcc-TiN seed layer has a 4-fold rotation axis along the $[00\bar{1}]_{fcc}$ direction. By the simultaneous appearance of the orientations "w3.1", "w3.2", "w3.3" and "w3.4", as seen in the SAED pattern in the sample FIB_3, the w-AlN crystallites create a quasi 4-fold rotational symmetry and adapt to the TiN seed layer. This is illustrated in Fig. 7.71.

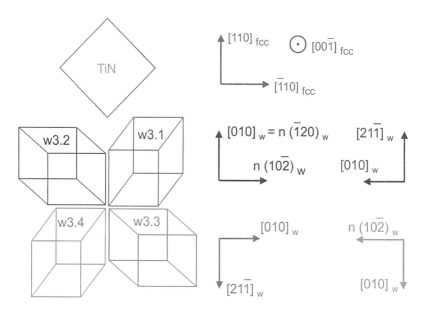

Fig. 7.71: Sketch of the fcc-TiN and w-AlN elementary cells showing the alignment of the orientation types "w3.1", "w3.2", "w3.3" and "w3.4" with respect to the $(00\bar{1})$ oriented fcc-TiN seed layer. The direction pointing out of the paper corresponds to the growth direction of the w-AlN layer.

The presence of the orientations "w3.1", "w3.2", "w3.3"and "w3.4", as seen in the sample FIB_3, was proven by the orientations "w4.1", "w4.2", "w4.3"and "w4.4" in the sample FIB_4. Regarding the orientations "w3.1", "w3.2", "w3.3" and "w3.4", it is obvious that the $\{120\}_w$ planes aligne parallel to the $\{110\}_{fcc}$ planes of the fcc-TiN seed layer (see Fig. 7.71). This was also observed for the w-AlN coating deposited onto the $(1\bar{1}0)_{fcc}$ oriented fcc-TiN seed layer (samples FIB_1, FIB_2). The same alignment of the lattice planes was found for the OR given in Eq. (7.19). Thus, also the w-AlN deposited onto the $(00\bar{1})$ oriented fcc-TiN seed layer indicates the preferred alignment of the lattice planes $(\bar{1}20)_w$ and $(220)_{fcc}$ which can be attributed to the small misfit of 3.6 % of the interplanar spacings of the $(\bar{1}20)_w$ and $(220)_{fcc}$ lattice planes.

7.4 Cr-Al-Si-N coatings deposited by UBM and HIPIMS

In this Section the effect of the ion bombardment on the microstructure formation in Cr-Al-Si-N coatings is presented. Within this work Cr-Al-Si-N coatings were deposited using unbalanced magnetron sputtering (UBM) and high power impulse magnetron sputtering (HIPIMS). The microstructure of these coatings is compared to the microstructure of Cr-Al-Si-N coatings that were deposited by cathodic arc evaporation (CAE). Prior to this study, the Cr-Al-Si-N coatings deposited by CAE were characterized in Refs. [10, 151]. Since the flux of the ionized coating species present in the plasma increases from UBM to HIPIMS to CAE, the microstructure formation was correlated to the ion bombardment during the deposition of Cr-Al-Si-N coatings. A part of the results of the study was published in Ref. [137].

7.4.1 Deposition of Cr-Al-Si-N coatings using UBM, HIPIMS and CAE

The **UBM** and **HIPIMS** depositions were done at the Nanotechnology Centre for PVD Research at the Sheffield Hallam University in Great Britain. A CSM-18 vacuum system from Kurt J. Lesker utilizing two planar unbalanced magnetrons with circular cathodes with a diameter of 75 mm was used for the coating deposition. The cathodes were furnished with targets of Cr and Al-Si. The Al-Si cathode contained 11 at.% Si. The cathodes were aligned confocally to the substrate so that the angle between the substrate normal and target normal enclose an angle of 32°. A schematic drawing of the deposition device is given in Fig. 7.72a and further details are given in Ref. [13].

Prior to the deposition of the **UBM** coatings, the substrates were etched in Ar plasma whilst operating the Cr cathode behind shutters. The substrate bias voltage during etching was -1000 V. The UBM deposition was carried out with both cathodes operating in dc mode. The [Cr] / ([Cr] + [Al] + [Si]) ratio in the samples was varied through the power ratio $P(Cr) = (P_{Cr}/(P_{Cr} + P_{Al-Si}))$ as shown in Table 7.5. The average total power of both targets ($P_{Cr} + P_{Al-Si}$) ranged between 0.75 and 0.80 kW.

The **HIPIMS** deposition was carried out utilising the Cr cathode in the HIPIMS mode and the Al-Si cathode in the DC mode, thus the deposition is actually a **hybrid HIPIMS / MS** deposition technology. For simplification it is often labelled as **HIPIMS** deposition in the following.

Prior to the deposition of the HIPIMS sample series, the substrates were etched by Cr and Ar ions produced in the HIPIMS Cr plasma. The substrate bias was -600 V during etching. In the HIPIMS process, a series of coatings with different chemical compositions was produced using the same set of the $P_{Al\text{-}Si}$ powers as in the case of the UBM sputtering (cf. Table 7.5 and Table 7.6). The average power P_{Cr} on the HIPIMS target was adjusted by varying the peak power on the Cr target. The pulse frequency and the pulse duration were kept constant at 300 Hz and 80 μs, respectively. In this study, Cr-Al-Si-N coatings were deposited by HIPIMS with similar [Cr] / ([Cr] + [Al] + [Si]) ratios as achieved in the UBM coatings. Taking into account the generally lower deposition rates in the HIPIMS process, the average power on the Cr HIPIMS cathode (P_{Cr}) was increased in comparison with the power applied to the Cr DC cathode during the UBM deposition.

Table 7.5: Power on the respective target, P_{Cr} and $P_{Al\text{-}Si}$, the power ratio, $P(Cr) = P_{Cr}/(P_{Cr} + P_{Al-Si})$, the relative Cr, Al and Si contents, [Cr] / Σ[X], [Al] / Σ[X] and [Si] / Σ[X] (where Σ[X] = ([Cr] + [Al] + [Si])), the [Si] / ([Al] + [Si]) ratio as determined by EPMA / WDS and the thickness and indentation hardness of individual Cr-Al-Si-N coatings deposited by UBM.

P_{Cr} (kW)	P_{Al-Si} (kW)	$P(Cr)$	$\frac{[Cr]}{\Sigma[X]}$	$\frac{[Al]}{\Sigma[X]}$	$\frac{[Si]}{\Sigma[X]}$	$\frac{[Si]}{[Al]+[Si]}$	Thickness (μm)	H_{IT} (GPa)
0.50	0.27	0.65	0.87	0.12	0.01	0.08	6.8 ± 0.4	16.4±0.7
0.37	0.42	0.47	0.75	0.23	0.02	0.08	5.3 ± 0.1	12.9±0.4
0.26	0.54	0.33	0.61	0.37	0.02	0.05	4.7 ± 0.3	20.8±0.5
0.15	0.61	0.20	0.44	0.52	0.04	0.07	3.8 ± 0.3	15.4±0.3
0.10	0.65	0.13	0.26	0.69	0.05	0.07	4.3 ± 0.1	16.5±0.2

Table 7.6: Power on the respective target, P_{Cr} and $P_{Al\text{-}Si}$, the power ratio, $P(Cr) = P_{Cr}/(P_{Cr} + P_{Al-Si})$, the relative Cr, Al and Si contents, [Cr] / Σ[X], [Al] / Σ[X] and [Si] / Σ[X] (where Σ[X] = ([Cr] + [Al] + [Si])), the [Si] / ([Al] + [Si]) ratio as determined by EPMA / WDS and the thickness and indentation hardness of individual Cr-Al-Si-N coatings deposited by HIPIMS.

P_{Cr} (kW)	P_{Al-Si} (kW)	$P(Cr)$	$\frac{[Cr]}{\Sigma[X]}$	$\frac{[Al]}{\Sigma[X]}$	$\frac{[Si]}{\Sigma[X]}$	$\frac{[Si]}{[Al]+[Si]}$	Thickness (μm)	H_{IT} (GPa)
0.71	0.28	0.72	0.80	0.19	0.01	0.05	3.4 ± 0.1	33.6±0.8
0.52	0.42	0.55	0.67	0.31	0.02	0.06	3.9 ± 0.1	26.3±0.8
0.36	0.54	0.40	0.50	0.47	0.03	0.06	4.2 ± 0.2	33.4±1.2
0.20	0.60	0.25	0.44	0.53	0.03	0.05	3.7 ± 0.3	23.4±0.5
0.14	0.64	0.18	0.27	0.68	0.05	0.07	4.5 ± 0.4	15.1±0.4

Both HIPIMS and UBM depositions were done in a mixed argon and nitrogen atmosphere at a total pressure of 0.36 Pa and at the nitrogen partial pressure of 0.16 Pa. The deposition time was 300 min, the temperature in the deposition apparatus 400°C and the bias voltage -75 V. As substrates mirror-polished cemented carbides containing hex-WC and 10 wt.% Co were used.

A plasma sampling energy-resolved mass spectrometer PSM003 from Hiden Analytical Ltd. was used to measure the ion energy distribution function by Hecimovic [233] for all five sets of the HIPIMS deposition parameters and for one UBM deposition run with $P(Cr) = 0.33$. The spectrometer was located at a distance of 17 cm from the target and enclosed an angle of 58° with the target normal [13]. Since the target spectrometer distance was comparable with the target-substrate distance, the plasma composition registered by the instrument was practically the same as the plasma composition in the vicinity of the substrates. The ion energy distribution function was measured in the time-averaged mode for the Cr^{1+}, Cr^{2+}, Al^{1+}, Ar^{1+}, N_2^{1+} and N^{1+} species. The relative amount of the respective ions in the plasma were deduced from the energy-integrated ion energy distribution function for each species [233].

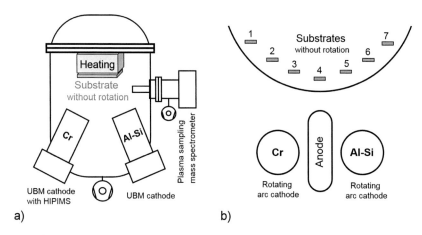

Fig. 7.72: Schematic drawing of the used deposition device CSM-18 from Kurt J. Lesker (a) and cross-section of the deposition apparatus π-80 from PLATIT (b).

The deposition of the Cr-Al-Si-N coatings using **CAE** was done by SHM Ltd. Šumperk in Czech Republic. The coatings were not analysed within this study. Within this study the results of the microstructure analysis of the CAE Cr-Al-Si-N coatings as published in Ref. [10] were compared with the microstructure of the UBM and HIPIMS coatings. The deposition of the CAE Cr-Al-Si-N coatings was performed in the deposition apparatus π-80 from PLATIT [234] that is equipped with two laterally rotating arc cathodes. As in the case of the UBM and HIPIMS coatings one cathode was made of Cr the other one was made of Al containing 11 at.% Si. The ion current on the Cr and Al-Si cathode was 80 A and 120 A, respectively. The deposition was done in nitrogen atmosphere at a pressure of 1.3 Pa and at a temperature of 450 °C. The bias voltage applied to the substrates was -75 V. The samples were not rotated during the deposition process.

Table 7.7: The relative Cr, Al and Si contents, [Cr] / Σ[X], [Al] / Σ[X] and [Si] / Σ[X] (where Σ[X] = ([Cr] + [Al] + [Si])), the [Si] / ([Al] + [Si]) ratio as determined by EPMA / WDS and the thickness and indentation hardness of individual CAE Cr-Al-Si-N coatings as taken from Ref. [10].

Sample	$\dfrac{[Cr]}{\Sigma[X]}$	$\dfrac{[Al]}{\Sigma[X]}$	$\dfrac{[Si]}{\Sigma[X]}$	$\dfrac{[Si]}{[Al]+[Si]}$	Thickness (μm)	H_{IT} (GPa)
1	0.91	0.08	0.01	0.11	4.5 ± 0.2	27.7 ± 0.7
2	0.84	0.15	0.01	0.06	6.3 ± 0.2	29.6 ± 1.2
3	0.69	0.28	0.03	0.10	7.9 ± 0.3	34.4 ± 1.6
4	0.52	0.43	0.05	0.10	8.7 ± 0.3	41.5 ± 0.5
5	0.40	0.52	0.08	0.13	8.6 ± 0.3	44.5 ± 1.7
6	0.24	0.65	0.10	0.13	7.5 ± 0.3	39.1 ± 1.3

7.4.2 Ionized species in the UBM and HIPIMS plasma

The relative contents of the individual ions in the plasma as measured by Hecimovic [233] using an energy-resolved mass spectrometer are summarized in Fig. 7.73b for one UBM deposition run with a power ratio $P(Cr)$ of 0.33 (see Table 7.5) and three hybrid HIPIMS / MS deposition runs with a power ratio $P(Cr)$ of 0.72, 0.40 and 0.25 (see Table 7.6). The HIPIMS peak current decreased with decreasing average power ratio $P(Cr)$ (Fig. 7.73a) during the hybrid HIPIMS / MS deposition runs that indicates a decreasing ionization degree. During the UBM depositions the HIPIMS peak current was zero.

The relative metal ion content is influenced on the one hand by a different degree of ionisation in the plasma and on the other hand by different proportions of the metallic species in the plasma. In order to correct the relative contents of the individual metal ions in the plasma with respect to the coating composition, the relative content of metal ions is divided by the respective metal ratio in the coating as determined by EMPA / WDX and plotted in Fig. 7.73c.

From Fig. 7.73 the influence of the HIPIMS power on the relative ion content is obvious. The relative ion content for the UBM deposition illustrates the case of no application of HIPIMS power to the targets. In case of the hybrid HIPIMS / MS deposition runs, only the Cr cathode was operated in HIPIMS mode. Thus, the increase of the power ratio $P(Cr)$ means concurrently an increase of the HIPIMS peak power.

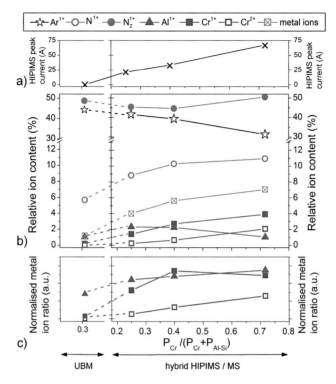

Fig. 7.73: HIPIMS peak current (a) and relative content of individual ions in the plasma as measured by energy resolved mass spectrometry during selected deposition runs (b) as well as relative content of the metal ions from figure (b) divided by the respective metal ratios [M]/([Cr]+[Al]+[Si]) with M = Cr, Al (c) for one UBM deposition and three hybrid HIPIMS / MS depositions. The results are based on Ref. [233].

For both deposition techniques, the majority of the ions are Ar^{1+} and N_2^{1+} stemming from the working and reactive gas. Both ions made up more than 80 % of the total ion content in the plasma. The highest relative amount of Ar^{1+} was observed in the UBM process, where the ionization of other species was low. With increasing HIPIMS power on the Cr target and thus with increasing $P(Cr)$ in the HIPIMS process, the relative amount of Ar^{1+} decreased steadily and the relative ion content of the other ions (mainly Cr^{1+} and Cr^{2+}) increased. The dominant ions in the UBM and HIPIMS depositions were N_2^{1+}. Starting from zero HIPIMS power in the UBM process, the relative ratio of this ion type decreased with increasing HIPIMS power up to $P(Cr) = 0.40$. As in the case of Ar^{1+}, the reason for this is the increasing ionisation of chromium atoms. For the highest HIPIMS power ($P(Cr) = 0.72$), a very high ratio of N_2^{1+} was observed, which is attributed to a high degree of ionization of the plasma and the additional ionisation of the nitrogen molecules by the high ionisation of the metallic species. The next highest concentration was found for the ionised dissociated nitrogen atoms N^{1+}. They represent 6 % of the ions in the UBM deposition. With increasing HIPIMS power their ratio increased to 11 % due to an increasing degree of ionization in the plasma.

The concentration of the metal ions were significantly lower than the concentrations of the gas ions. In Fig. 7.73 only the relative concentrations of the Cr^{1+}, Cr^{2+} and Al^{1+} are shown because the relative concentrations of Si^{1+}, Al^{2+} and other ions were below the detection limit. Starting from zero HIPIMS power in the UBM process and increasing the HIPIMS power in the hybrid HIPIMS / MS mode, the relative metal ion content increases (see Fig. 7.73). This illustrates the higher degree of ionisation of the sputtered material during the transition from the UBM to the HIPIMS process. Regarding metal ion fractions that were normalised to the coating composition in Fig. 7.73c, it is obvious that the fraction of each metal ion species was higher in the HIPIMS mode than in the UBM deposition. Although the HIPIMS discharge was applied only to the Cr target, a significant fraction of Al^{1+} ions was produced by the Cr HPIMS discharge as compared to the UBM process. Their fraction increased only slightly with increasing HIPIMS power (see Fig. 7.73c). In case of Cr^{1+} ions, the ion content normalized to the coatings composition reached saturation at medium HIPIMS power. In contrast to that, the normalized ion fraction of the Cr^{2+} increased continuously with increasing HIPIMS power. In general, the increasing HIPIMS power enhanced the dissociation and ionisation of nitrogen and the ionisation of metallic species, especially the double ionisation of chromium.

The characteristics of the UBM and HIPIMS deposition processes influence the relation between the relative power of the Cr cathode ($P(Cr)$) and the $[Cr] / ([Cr] + [Al] + [Si])$ ratio (Fig. 7.74a) as well as the specific deposition rate (Fig. 7.74b) of the Cr-Al-Si-N coatings. For both deposition methods, the $[Cr] / ([Cr] + [Al] + [Si])$ ratio increases with increasing $P(Cr)$. In case of the HIPIMS coatings the increase is linear whereas the UBM coatings deviate from a linear dependence. In general the UBM coatings are characterized by a higher $[Cr] / ([Cr] + [Al] + [Si])$ ratio than the HIPIMS coatings if the same mean $P(Cr)$ is considered (see Fig. 7.74a). This is attributed to a higher efficiency of the Cr deposition from the UBM cathode than from the HIPIMS cathode especially for $P(Cr) < 0.5$. For $P(Cr) > 0.5$, the efficiency of the Cr incorporation into the HIPIMS coatings is apparently improved since the difference in the Cr concentration between the UBM and HIPIMS coatings decreased slightly (see Fig. 7.74a).

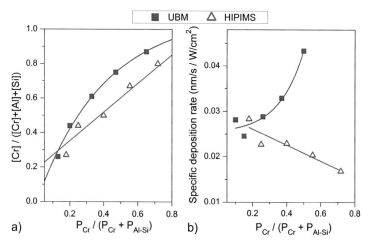

Fig. 7.74: Chromium ratio (a) and specific deposition rate (b) of the UBM and HIPIMS coatings as function of $P(Cr) = P_{Cr}/(P_{Cr} + P_{Al-Si})$.

The deposition rates for the Cr-Al-Si-N coatings deposited by UBM and HIPIMS plotted in Fig. 7.74b were calculated from the thickness of the coatings divided by the deposition time and related to the average power density applied to the targets (W/cm^2). In case of the UBM deposition, the specific rate increased steeply with increasing $P(Cr)$. In contrast to that the specific deposition rate in the HIPIMS deposition process decreased nearly linearly with increasing $P(Cr)$ and with increasing HIPIMS power on the chromium cathode. This decrease reflects the higher re-sputtering of the deposited species, which is attributed on the one hand to a higher content of metal ions in the plasma and on the other to an increase of multiple charged metal ions, e.g. Cr^{2+} as shown in Fig. 7.73.

7.4.3 Phase composition of the Cr-Al-Si-N coatings deposited by UBM and HIPIMS and crystal anisotropy of the lattice deformation

The GAXRD patterns in Fig. 7.75 showed that fcc-(Cr,Al,Si)N phase was the only crystalline phase in the Cr-Al-Si-N coatings of both deposition types with $[Cr] / ([Cr] + [Al] + [Si]) \geq 0.44$. At higher Al contents the w-AlN phase became the dominant crystalline phase in the coatings deposited by UBM and HIPIMS. According to Ref. [10], the formation of the w-AlN phase in the Cr-Al-Si-N coatings deposited by CAE was already observed at $[Cr] / ([Cr] + [Al] + [Si]) \leq 0.40$.

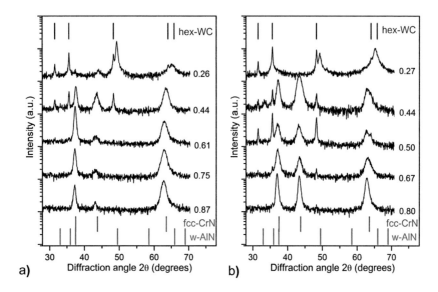

Fig. 7.75: Parts of the GAXRD patterns measured for Cr-Al-Si-N coatings deposited by UBM (a) and HIPIMS (b). The positions of fcc-CrN and w-AlN are labelled at the bottom of the figure and the positions of hex-WC originating from the cemented carbide substrate are shown at the top of the figure. The numbers at the right of each diffraction pattern correspond to the [Cr] / ([Cr]+[Al]+[Si]) ratio in the coatings.

Detailed analysis of the GAXRD patterns with the aid of X-ray line profile fitting revealed a large anisotropy of the elastic lattice deformation in the fcc-phase. The kind and the extent of this anisotropy can be recognized from the scatter of the measured lattice parameters a_ψ^{hkl} when they are plotted vs. $\sin^2\psi$. A comparison of the a_ψ^{hkl} vs. $\sin^2\psi$ plots for the Cr-Al-Si-N coatings deposited by UBM and HIPIMS, which contain high and low chromium contents, are given in Fig. 7.76. For the $Cr_{0.84}Al_{0.15}Si_{0.01}N$ coating deposited by CAE the dependence of a_ψ^{hkl} vs. $\sin^2\psi$ is shown additionally in Fig. 7.76a as taken from Ref. [137] The symbols in Fig. 7.76 correspond to the measured lattice parameters. The dependence of a_ψ^{hkl} as function of $\sin^2\psi$ was described by Eq. (6.27) and is shown as broken line for the five coatings in Fig. 7.76.

From the a_ψ^{hkl} vs. $\sin^2\psi$ plots in Fig. 7.76 it is visible that the anisotropy of the elastic lattice deformation of the Cr-Al-Si-N coatings deposited by UBM differs from the anisotropy in the coatings deposited by HIPIMS and CAE. The anisotropy of the elastic lattice deformation in the fcc-phase present in the $Cr_{0.80}Al_{0.19}Si_{0.02}N$ coating deposited by HIPIMS is typical for transition metal nitrides with B1-NaCl crystal structure like e.g. ZrN, TiN, HfN [143, 148]. In

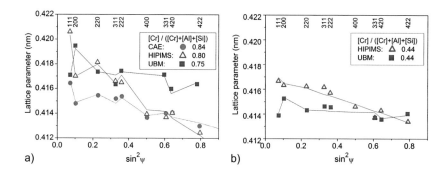

Fig. 7.76: Dependence of the lattice parameters of the fcc phase on $\sin^2\psi$ of the HIPIMS (Δ), UBM ($\blacksquare$) and CAE ($\bullet$) Cr-Al-Si-N coatings containing high (a) and low (b) $[Cr]/([Cr]+[Al]+[Si])$ ratio. The respective diffraction indices are given at the top of the plot. The broken lines were calculated according to Eq. (6.27). The results of the Cr-Al-Si-N coating deposited by CAE is taken from Refs. [10, 137].

these nitrides, the easy deformation direction is $\langle 111 \rangle$. Thus the lattice parameter a^{111} is larger than a^{200} for coatings under compressive stress. Such an anisotropy of the elastic lattice deformation was observed also in different ternary and quaternary transition metal nitride coatings using CAE, e.g. $Ti_{1-x}Al_xN$ coatings for $x < 0.28$ [215], $Cr_{1-x-y}Al_xSi_yN$ [10], $Cr_{1-x}Al_xN$, $Zr_{1-x}Al_xN$ and $Ti_{1-x-y}Al_xSi_yN$ [147].

Regarding Fig. 7.76, the degree of the anisotropy of the lattice deformation decreased with increasing aluminium content in the HIPIMS coatings. This can be described by the anisotropy factor A that was calculated for the fcc-(Cr,Al,Si)N phase using Eq. (6.42) according to the

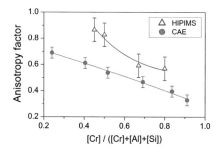

Fig. 7.77: Anisotropy factors of the fcc-(Cr,Al,Si)N phase as determined by Eq. (6.42) from the dependence of the lattice parameters on $\sin^2\psi$ in the Cr-Al-Si-N coatings with different $[Cr]/([Cr]+[Al]+[Si])$ ratios. The values for the Cr-Al-Si-N coatings deposited by CAE are taken from Ref. [137].

procedure described in Section 6.2.1.2 (see page 37). They are plotted in Fig. 7.77 for the Cr-Al-Si-N coatings that were deposited by CAE and HIPIMS. In both coating types the anisotropy factors were below 1. The CAE and HIPIMS coatings with the highest Cr content were characterized by the largest anisotropy. With increasing Al content in the CAE and HIPIMS coatings the extent of the anisotropy decreased. From the point of the anisotropy of the elastic lattice deformation, the Cr-Al-Si-N coatings deposited at high HIPIMS peak power (high $[Cr] / ([Cr] + [Al] + [Si])$ ratio) are similar to the coatings deposited by CAE. With decreasing HIPIMS peak power and thus with decreasing $[Cr] / ([Cr] + [Al] + [Si])$ ratio (≤ 0.50), the Cr-Al-Si-N coatings deposited by HIPIMS deviated from the behaviour of the coatings deposited by CAE. The coatings deposited by CAE are characterized by a certain anisotropy of the lattice deformation up to high Al content, whereas nearly no anisotropy of the lattice deformation of the fcc-phase was detected in the HIPIMS coatings with $[Cr] / ([Cr] + [Al] + [Si]) \leq 0.44$. In the Cr-Al-Si-N coatings deposited by UBM, the opposite anisotropy of the lattice deformation, as indicated by $a^{111} < a^{200}$ for compressive residual stress, was found. The anisotropy factor ranged between 2 and 3.

7.4.4 Stress-free lattice parameter of the fcc-(Cr,Al,Si)N phase

The analysis of the stress-free lattice parameters of the fcc-(Cr,Al,Si)N phase present in the Cr-Al-Si-N coatings deposited by UBM and HIPIMS and the comparison with the stress-free lattice parameters found in the Cr-Al-Si-N coatings deposited by CAE [10] indicated that the lattice parameter is affected both by the chemical composition and by the deposition mode.

Regarding the stress-free lattice parameter of the fcc-(Cr,Al,Si)N phase found in the Cr-Al-Si-N coatings deposited by UBM, the stress-free lattice parameter decreased linearly with decreasing $[Cr] / ([Cr] + [Al] + [Si])$ ratio in the coatings. The shrinkage of the lattice parameter of the fcc-phase with increasing Al content is also visible in the Cr-Al-(Si)-N coatings deposited by CAE (see grey symbols in Fig. 7.78) [109]. This effect is also reported by many other authors [113, 115, 116, 120-122].

For the Cr-Al-(Si)-N films deposited by CAE, the influence of the incorporation of Al and Si into the fcc-$Cr_{1-x-y}Al_xSi_yN$ phase on the lattice parameter was described in Ref. [109] by Eqs.(5.1) and (5.2), respectively.

The dependence of Eqs. (5.1) and (5.2) is plotted by a grey line in Fig. 7.78 for the Cr-Al-N coatings (straight line) and by a broken line for the Cr-Al-Si-N coatings, respectively. The departure of the stress-free lattice parameter of the fcc-phase in the Cr-Al-N coatings from the straight line for [Cr] / ([Cr] + [Al]) ≤ 0.45 indicates the phase segregation and formation of w-AlN [109]. In case of the Cr-Al-Si-N coatings deposited by CAE, the lattice parameter is increased as compared to the Si-free CAE coatings which indicates the inflation of the lattice

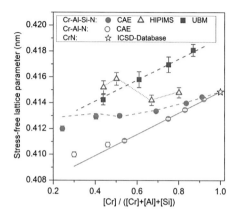

Fig. 7.78: Dependence of the stress-free lattice parameter on the Cr ratio in the Cr-Al-Si-N coatings deposited by HIPIMS and UMBM. The stress-free lattice parameters of the fcc phase in the Cr-Al-(Si)-N coatings deposited by CAE were adopted from Ref. [10] and are shown for comparison. The lattice parameter of fcc-CrN was taken from the ICSD database [235].

parameter due to the incorporation of Si in the fcc-phase.

The stress-free lattice parameters observed in the UBM coatings are much larger than the ones in the CAE coatings with similar Si and Al contents. Furthermore, some of them exceed the lattice parameter of fcc-CrN [235]. On the other hand the lattice parameter depends linearly on the [Cr] / ([Cr] + [Al] + [Si]) ratio and the slope of this linear dependence is similar to the slope observed for the Si-free Cr-Al-N coatings (see Eq. (5.1)).

Three possible reasons for the increased lattice parameter of the fcc-phase in the Cr-Al-Si-N coatings deposited by UBM were considered.

(i)   incorporation of argon or additional nitrogen into the volume of the elementary cell of fcc-(Cr,Al,Si)N [207]

(ii)   increasing Si content in fcc-$Cr_{1-x-y}Al_xSi_yN$ [109]

(iii) the ordering of the magnetic moments [139]

Using energy dispersive spectroscopy in STEM, no evidence of argon incorporation into the coatings deposited by UBM was found. Thus, the effect of the incorporation of argon into the elementary cell of fcc-(Cr,Al,Si)N was neglected. Furthermore, EMPA / WDS did not indicate differences in the nitrogen content in the coatings deposited by UBM and HIPIMS. Hence, it is concluded that the expansion of the fcc elementary cell especially in the coatings deposited by UBM is not related to excess nitrogen content.

In general, silicon embedded in the elementary cell of fcc-(Cr,Al,Si)N expands the lattice parameter [109]. On the one hand, the coatings deposited by UBM contain a higher $[Si] / ([Al] + [Si])$ ratio than the coatings deposited by HIPIMS, but the $[Si] / ([Al] + [Si])$ ratio is higher in the coatings deposited by CAE (see Table 7.5, Table 7.6 and Table 7.7). This indicates that a possible presence of silicon in fcc-(Cr,Al,Si)N in the UBM coatings cannot fully explain the observed expansion of the stress-free lattice parameter. On the other hand, if the dependence of the lattice parameter of the fcc phase on the $[Cr] / ([Cr] + [Al] + [Si])$ ratio is compared, a similar slope of the Si-free Cr-Al-N coatings deposited by CAE (see Fig. 7.78 and Eq. (5.2)) and the Cr-Al-Si-N coatings deposited by UBM coatings is evident. This suggests that the silicon present in the coatings deposited by UBM is probably not incorporated in the fcc-phase.

Due to these reasons, it is assumed that the expansion of the elementary cell in the coatings deposited by UBM originates from the ordering of the magnetic moments. Considering the calculations of Filippetti *et al.* the ferromagnetic and the antiferromagnetic ordering leads to an increase of the lattice parameter if the fcc structure is retained whereby from the energetically point of view the antiferromagnetic ordering is favoured over the ferromagnetic ordering [139] (see Section 5.2).

Furthermore, the hypothesis of the magnetic ordering was supported by *ab initio* calculations in Ref. [137] which revealed an anisotropy factor of the elastic lattice deformation with $A > 1$ for the antiferromagnetic ordering in the pseudo-cubic elementary cell of CrN that is inverse as compared to the non-magnetic structure of fcc-CrN with $A < 1$. The inverse elastic anisotropy with $A > 1$ was also observed in the Cr-Al-Si-N coatings deposited by UBM as shown in Fig. 7.76.

The lattice parameter of the fcc-phase present in the Cr-Al-Si-N coatings deposited by the HIPIMS process ranges between the lattice parameters observed in the coatings deposited by UBM and CAE. It is obviously controlled by the HIPIMS power, since for a high HIPIMS power and an average Cr content of $[Cr] / ([Cr] + [Al] + [Si]) \geq 0.67$ the stress-free lattice parameter approaches the values observed in the coatings deposited by CAE. At low HIPIMS power and thus lower average $[Cr] / ([Cr] + [Al] + [Si])$ ratio of 0.50 and 0.44, the stress-free lattice parameter of both coatings increases and approaches the values observed in the Cr-Al-Si-N coatings deposited by UBM. Since in both coatings no indication for the presence of the wurtzite phase was apparent, this increase of the stress-free lattice parameter is attributed to the lower influence of the HIPIMS power rather than to the formation of wurtzite phase. The reduced impact of the HIPIMS power in the coatings with an average Cr content of $[Cr]/([Cr]+[Al]+[Si]) \leq 0.50$ means that the characteristics of the hybrid HIPIMS / MS deposition mode approaches the properties of the UBM process as shown in Fig. 7.73 and Section 7.4.2. This will be discussed in the next Section 7.4.5.

7.4.5 Influence of the deposition mode on the microstructure of Cr-Al-Si-N coatings

The analysis of the broadening of the X-ray diffraction lines of the fcc phase in the Cr-Al-Si-N coatings deposited by UBM and HIPIMS / MS revealed partial coherence of the nanocrystallites [200] having a small mutual misorientation ($< 2°$) similar to TM-Al-(Si)-N coatings (TM = Ti, Cr, Zr) deposited by CAE [57, 92, 201]. The crystallite and cluster sizes were determined by the procedure described in Ref. [200] and are plotted in Fig. 7.79 for the Cr-Al-Si-N coatings deposited by UBM and hybrid HIPIMS / MS with different $[Cr]/([Cr]+[Al]+[Si])$ ratios. The values of the Cr-Al-Si-N coatings deposited by CAE that were adopted from Ref. [10] are plotted for comparison Fig. 7.79. The crystallite size decreased continuously from ~ 11 to 4 nm with decreasing $[Cr]/([Cr]+[Al]+[Si])$ ratio in the coatings deposited by CAE, whereas the cluster size was in the range of ~ 50 nm [10]. The decrease of the crystallite size with increasing Al and Si content in the coatings deposited by CAE was attributed to the segregation of Si_3N_4 or AlN to the boundaries of the nanocrystallites [10].

The cluster size in the coatings deposited by UBM was in the range of ~ 20 nm and the crystallite size was ~ 4 nm and practically independent of the coating composition. In case of the coatings deposited by the hybrid HIPIMS / MS process with $[Cr]/([Cr]+[Al]+[Si]) \leq 0.50$, which

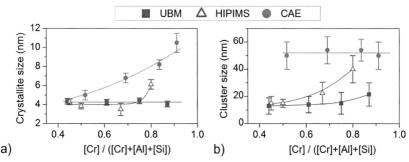

Fig. 7.79: Crystallite size (a) and cluster size (b) of the fcc-(Cr,Al,Si)N phase present in the coatings deposited by UBM, HIPIMS and CAE as function of the [Cr] / ([Cr]+[Al]+[Si]) ratio. The values of the Cr-Al-Si-N coating deposited by CAE were adopted from Ref. [10].

means a low HIPIMS power, the cluster size corresponds to the size in the coatings deposited by UBM. With increasing HIPIMS power the cluster size increased to ~ 40 nm and approached the values in the coatings deposited by CAE.

The differences in the cluster size in the coatings deposited by the three deposition techniques can be attributed to the differences in the degree of the plasma ionization especially the different density of film forming ions as well as the energy delivered to the growing film. In case of the sputtered coatings (UBM), the ion flux is dominated by species from the working and reactive gas and just a minor part of the sputtered material is ionized (see Fig. 7.73) [13]. Most of the sputtered species reach the substrate as neutrals with a low kinetic energy of just a few eV. The neutral species are not influenced by the applied bias voltage of -75 V. With increasing HIPIMS power, the ion to neutral flux of the film forming species reaching the substrate increases. According to the literature, most of the ions in a HIPIMS plasma are single charged but also a significant fraction of 2+ charge states is present [18]. This agrees with the relative content of the individual ions which were measured by Hecimovic [233] during the hybrid HIPIMS / MS deposition of the Cr-Al-Si-N coatings and which are shown in Fig. 7.73. The ion energy distribution function is characterized by a low energy peak and a high energy tail [13, 18]. For example in the case of Cr^{1+} ions observed in a HIPIMS plasma, the low energy ions had a peak at 3 eV and the energy tail reached up to 60 eV depending on the time after the discharge ignition [236]. The energy of the ions arriving at the substrate can be controlled by the applied substrate bias voltage of -75 V.

The low energy ion irradiation is known to enhance the mobility of the adatoms resulting in denser coatings, as shown for instance in Refs. [237, 238], especially if the energy is below the subplantation threshold [69]. Thus, the increased ion flux (bombardment) during HIPIMS deposition of the Cr-Al-Si-N coatings with high HIPIMS power led to a higher surface mobility of the adatoms as compared to the coatings deposited by UBM [18]. The higher mobility of the ad-atoms led obviously to the growth of larger clusters with increasing HIPIMS power that approached the cluster size in the Cr-Al-Si-N coatings deposited by CAE (see Fig. 7.79b).

The smaller crystallite size in the Cr-Al-Si-N coatings deposited by UBM and HIPIMS as compared to the crystallite size in the Cr-Al-Si-N coatings deposited by CAE is attributed to the segregation of silicon nitrides to the crystallite boundaries. In contrast to these coatings, the partial incorporation of Si into fcc-(Cr,Al,Si)N was observed in the Cr-Al-Si-N coatings deposited by CAE [10]. However, at high Al and Si contents in the CAE coatings with $[Cr]/([Cr]+[Al]+[Si]) \leq 0.52$ the segregation of silicon nitride resulted in similar crystallite sizes of ~ 4 - 5 nm like in the Cr-Al-Si-N coatings deposited by UBM and HIPIMS.

The influence of the ion irradiation depending on the used deposition mode is also indicated by the stress-free lattice parameter of the fcc phase as shown in Fig. 7.76. As shown in Section 7.4.4, the Cr-Al-Si-N coatings deposited by UBM were characterized by an unusual increase of the stress-free lattice parameter, which can originate from the ordering of the magnetic moments of chromium [139].

The stress-free lattice parameter in the coatings deposited by CAE decreased continuously with increasing Al content; no unexpected inflation of the elementary cell was observed [10]. The CAE deposition process is characterized by a high ion flux arriving at the substrate which is provided by a fully ionized plasma of the arc process [19]. The ions have a high natural kinetic energy E_{i0} (see Eqs.(2.4) and (2.5)) and they are multiply charged [18, 21]. Thus, the CAE deposition is an energetic deposition process, in which each ion delivers significant energy to the growing film. Due to subplantation of high energetic ions, defects are added to the film which cannot anneal completely [237]. The presence of a high fraction of point defects in the Cr-Al-Si-N coatings deposited by CAE, as compared to the coatings deposited by UBM, will probably disturb the magnetic ordering. Thus, it is expected that no magnetic ordering will appear in the Cr-Al-Si-N coatings deposited by CAE.

The lattice parameters observed in the coatings deposited by the hybrid HIPIMS / MS process approached at high HIPIMS power (coatings with $[Cr]/([Cr]+[Al]+[Si]) \geq 0.67$) the lattice parameters in the CAE coatings and at lower HIPIMS powers (coatings with $[Cr]/([Cr]+[Al]+[Si]) \leq 0.50$) the lattice parameters in the UBM coatings. Due to the higher ion to neutral flux of the film forming species at high HIPIMS powers, more energy is delivered to the growing film as compared to low HIPIMS powers. Thus, obviously a higher fraction of point defects that could disturb the magnetic ordering is formed at high HIPIMS power. This effect is apparently reduced with decreasing HIPIMS power in the Cr-Al-Si-N coatings with $[Cr]/([Cr]+[Al]+[Si]) \leq 0.50$ as indicated by the approach of the lattice parameter to the lattice parameters observed in the UBM coatings.

The influence of the deposition mode on the microstructure and properties is also apparent from the macroscopic lattice strain and the hardness of the coating. The macroscopic lattice strain of the fcc-phase is plotted in Fig. 7.80a. All coatings are characterized by a compressive stress state. In order to see if the compressive stress correlates with the thickness of the coatings, the thickness is shown in Fig. 7.80b. A correlation of the lattice strain and the thickness was reported for the Cr-Al-Si-N coatings deposited by CAE because with increasing Al content the macroscopic lattice strain and the thickness of the coatings increased [151]. In case of the coatings deposited by the UBM and the hybrid HIPIMS / MS process no correlation between strain and thickness could be asserted.

But the thickness of these coatings reflects the influence of the deposition mode. Although the deposition time for the coatings deposited by the UBM and the hybrid HIPIMS / MS process was equal, the Cr-Al-Si-N coatings deposited by the hybrid HIPIMS / MS process at a high HIPIMS power with an average Cr content of $[Cr] / ([Cr] + [Al] + [Si]) \geq 0.67$ are thinner than the UBM coatings with a similar chemical composition (see Fig. 7.80b). This effect is attributed to the lower deposition rate during the hybrid HIPIMS / MS process (see Fig. 7.74b) caused by the higher re-sputtering rate of deposited species due to a higher content of metal ions and the presence of double charged metal ions (see Fig. 7.73) at high HIPIMS powers. At lower HIPIMS power in the coatings with $[Cr] / ([Cr] + [Al] + [Si]) \leq 0.50$, the thickness resembles the thickness of the UBM coatings.

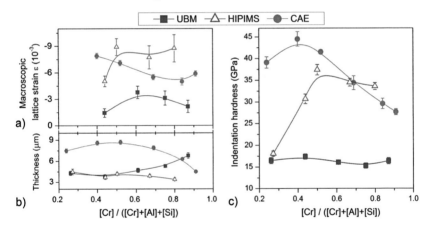

Fig. 7.80: Macroscopic lattice strain of the fcc-phase (a), coating thickness (b) and indentation hardness (c) of the Cr-Al-Si-N coatings deposited by UBM, hybrid HIPIMS / MS and CAE as function of the [Cr] / ([Cr]+[Al]+[Si]) ratio. The values of the Cr-Al-Si-N coating deposited by CAE were adopted from Ref. [151].

Regarding the macroscopic lattice strain, the coatings deposited by UBM are characterized by a significant lower strain as compared to the coatings deposited by CAE and the hybrid HIPIMS / MS process. Furthermore, the hardness of these coatings was significantly lower than in the other coatings (see Fig. 7.80c) if the composition range with $[Cr] / ([Cr] + [Al] + [Si]) > 0.30$ is considered where the fcc phase was the dominating phase. The low macroscopic lattice strain and hardness of the Cr-Al-Si-N coatings deposited by UBM is attributed to a low ionisation degree of the film forming species which resulted in a lower concentration of lattice defects which contributed to an increase of the compressive stress. Additionally, a lower mobility of the adatoms during UBM deposition as compared to the hybrid HIPIMS / MS deposition contributes to less dense UBM coatings, as reported for instance in Refs. [237, 238], and results in a low hardness.

A high ionisation degree of the film forming species and the presence of a significant fraction of double charged Cr^{2+} ions during the hybrid HIPIMS / MS process (see Fig. 7.73) at high HIPIMS power resulted in denser coatings and in the formation of ion induced defects that resulted in an increase of the strain and hardness.

Thus, at high HIPIMS powers the Cr-Al-Si-N coatings with $[Cr] / ([Cr] + [Al] + [Si]) \geq 0.67$ resemble the coatings deposited by CAE, which is apparent from a similar degree and kind of elastic anisotropy (Fig. 7.76 and Fig. 7.76), similar stress-free lattice parameter of the fcc-phase (Fig. 7.78), larger cluster size (Fig. 7.79), higher macroscopic lattice strain and hardness (Fig. 7.80).

With decreasing HIPIMS power and hence with a lower ionisation degree of the film forming species and a lower fraction of double charged Cr^{2+} ions as compared to high HIPIMS power, the microstructure approaches the microstructure of the Cr-Al-Si-N coatings deposited by UBM. This is especially obvious from the similar crystallite and cluster size (Fig. 7.79) and the similar stress-free lattice parameters of the fcc-phase (Fig. 7.78) for the HIPIMS coatings with $[Cr] / ([Cr] + [Al] + [Si]) \leq 0.50$ as compared to the coatings deposited by UBM. Also the reduction of the macroscopic lattice strain and hardness (Fig. 7.80) for the HIPIMS coatings with $[Cr] / ([Cr] + [Al] + [Si]) \leq 0.44$ indicate the reduced influence of the HIPIMS power on the microstructure.

8   Conclusions

In this work the influence of an energetic treatment using ion bombardment during the deposition of transition metal based nitride coatings on their microstructure and hardness was investigated in order to reveal the potential of energetic particle bombardment for defect engineering and material design of hard coatings. The study comprised the description of the correlation between the energetic treatment, microstructure, high-temperature stability and hardness of three groups of industrially relevant hard coatings: Ti-Al-N monolithic coatings, Ti-Al-N / Al-Ti-Ru-N multilayer coatings and Cr-Al-Si-N monolithic coatings. Detailed microstructure analysis was carried out using electron probe microanalysis with wavelength-dispersive spectroscopy, glow discharge optical emission spectroscopy, X-ray diffraction, transmission electron microscopy, energy dispersive spectroscopy and electron energy loss spectroscopy. The microstructure was described in the as-deposited state and partially *in situ* during and *ex situ* after thermal treatment up to 950°C. The obtained microstructure parameters were correlated with the indentation hardness of the coatings.

$Ti_{1-x}Al_xN$ coatings with four different Al contents in the range of $0.38 \leq x \leq 0.62$, which were deposited at four different bias voltages (U_B = -20 V, -40 V, -80 V and -120 V) using cathodic arc evaporation, were analyzed with regard to their microstructure in the as-deposited state and in the course of a thermal treatment. It was shown that the as-deposited microstructure of these coatings can be designed via the aluminium content and the bias voltage since both parameters influence the phase composition, the size of the fcc-(Ti,Al)N nanocrystallites and the residual stress and thus affect the hardness of the coatings.

The increase of the bias voltage resulted in an increase of the density of lattice defects, especially point defects created by a peeing process as well as the creation of local fluctuations of the Al and Ti concentration within (Ti,Al)N and an increase of the compressive stress. This behaviour can be attributed to a reduced mobility of the ad-atoms due to the high energetic ion bombardment at high bias voltages.

The $Ti_{1-x}Al_xN$ coatings deposited at a bias voltage of -20 V were nearly stress-free whereas the application of a bias voltage of -40 V led to the formation of compressive stress. In both cases the metastable fcc-(Ti,Al)N phase formed as a single phase in the $Ti_{1-x}Al_xN$ coatings with Al concentrations up to $x = 0.47$. This demonstrated a relative uniform distribution of Al and Ti in the Ti-Al-N coatings deposited at low bias voltages (U_B = -20 V, -40 V).

The application of high bias voltages (U_B = -80 V, -120 V) gave rise to an increasing defect density and resulted in a significant increase of compressive stress as compared to the coatings deposited at a bias voltage of -40 V. In the $Ti_{1-x}Al_xN$ coatings deposited at high bias voltages, a spatially non-uniform distribution of Ti and Al in (Ti,Al)N was found since these coatings contained the fcc-(Ti,Al)N phase as major phase and a low amount of the Al-rich fcc-(Al,Ti)N phase. The presence of a high compressive stress in the $Ti_{1-x}Al_xN$ coatings deposited at high bias voltages ($U_B = -80$ V, -120 V) helped to stabilize the metastable Al-rich fcc-(Al,Ti)N phase. The segregation of the Al-rich fcc-(Al,Ti)N phase resulted in a further fragmentation of the fcc-(Ti,Al)N clusters and led to the reduction of the crystallite size as compared to the single phase $Ti_{1-x}Al_xN$ coatings with $x \leq 0.47$ that were deposited at low bias voltages ($U_B = -20$ V, -40 V). Due to the segregated Al-rich fcc-(Al,Ti)N phase, the small crystallites and the high compressive stress, a hardness enhancement of the $Ti_{1-x}Al_xN$ coatings which were deposited at high bias voltages was achieved in comparison to the coatings deposited at low bias voltages.

The application of high bias voltages suppressed the formation of thermodynamically stable w-AlN with increasing Al content. This was especially obvious in the $Ti_{0.38}Al_{0.62}N$ coatings. Low volume fractions of w-AlN were found in the $Ti_{0.38}Al_{0.62}N$ coatings that were deposited at high bias voltages ($U_B = -80$ V, -120 V) whereas the wurtzite phase was the major phase in the $Ti_{0.38}Al_{0.62}N$ coatings deposited at low bias voltages ($U_B = -20$ V, -40 V). The formation of low volume fractions of w-AlN was observed already at Al concentrations of $x = 0.56$ in the $Ti_{1-x}Al_xN$ coatings that were deposited at low bias voltages ($U_B = -20$ V, -40 V). The indentation hardness of these coatings increased as compared to the single phase fcc-(Ti,Al)N coatings with $x \leq 0.47$ without increasing strongly the compressive residual stress. This supports the findings that the formation of w-AlN is not necessarily detrimental to the hardness of $Ti_{1-x}Al_xN$ coatings as it was often considered in the past. According to the results of this work, it is proposed that the hardness of $Ti_{1-x}Al_xN$ based coatings can be increased if the volume fraction of w-AlN is limited. A volume fraction of approx. 15 vol.% was still beneficial for the hardness. The influence of the wurtzite phase exceeding a critical volume fraction was obvious in the $Ti_{0.38}Al_{0.62}N$ coatings which were deposited at low bias voltages ($U_B = -20$ V, -40 V) and which were characterized by a hardness reduction.

The hardness enhancement with the occurrence of low volume fractions of w-AlN is attributed to the formation of partially coherent interfaces between fcc-(Ti,Al)N and w-AlN which introduce lattice strains. In the $Ti_{0.44}Al_{0.56}N$ coating deposited at $U_B = -40$ V the orientation relationship characterized by $(1\bar{1}0)_{fcc} \parallel (0\bar{1}0)_w$ (or $(\bar{1}\bar{1}0)_{fcc} \parallel (\bar{2}10)_w$) and $[001]_{fcc} \parallel [001]_w$ could be identified using HRTEM / FFT. This orientation relationship was found in two further Ti-Al-N based coatings.

In order to investigate such interfaces in more details, w-AlN was grown heteroepitaxially on $(1\bar{1}0)$ and $(00\bar{1})$ oriented fcc-TiN seed layers. The analysis of the local orientation relationship between the fcc-TiN seed layer and the columnar w-AlN grains demonstrated the formation of several possible orientation relationships for the formation of (partially) coherent TiN / AlN interfaces. Still, the experiments proved the preferential alignment of the $\{\bar{1}20\}_w$ planes parallel to the $\{110\}_{fcc}$ planes similar to the above mentioned orientation relationship found in industrially relevant Ti-Al-N based coatings. This is attributed to the small misfit of 3.6 % in the interplanar spacings of the $(220)_{fcc}$ and $(\bar{1}20)_w$ lattice planes.

The microstructure analysis of industrially relevant Ti-Al-N / Al-Ti-(Ru)-N multilayer coatings that were deposited at four different bias voltages ($U_B = -20$ V, -40 V, -60 V and -80 V) with three different Ru concentrations showed that the doping of the multilayers with Ru had neither a remarkable effect on the microstructure in the as-deposited state nor on the microstructure after thermal treatment in argon atmosphere. The microstructure and hardness is governed by the applied bias voltage during deposition using CAE. The phase analysis of the Ti-Al-N / Al-Ti-(Ru)-N multilayer coatings revealed the fcc-(Ti,Al)N phase as major phase and traces of w-AlN in all coatings. Still, no detrimental impact on the hardness due to the presence of traces of w-AlN was found. Like in the case of the monolithic $Ti_{1-x}Al_xN$ coatings, the segregation of an Al-rich fcc-(Al,Ti)N phase was observed in the coatings deposited at high bias voltages ($U_B = -60$ V, -80 V) which indicated local fluctuations of the Ti and Al concentration in (Ti,Al)N. In combination with a rising defect density due to the increased ion energy of the film forming species, an increase of the compressive stress was caused and an enhancement of the indentation hardness was obtained.

The bias voltage applied during the deposition of the monolithic $Ti_{1-x}Al_xN$ coatings and Ti-Al-N / Al-Ti-(Ru)-N multilayer coatings influenced also their thermal stability during annealing. During annealing of the Ti-Al-N based coatings, the metastable fcc-(Ti,Al)N phase decomposed in the temperature range of $850 - 950$ °C. The Ti-rich fcc-(Ti,Al)N and Al-rich fcc-(Al,Ti)N as well as wurtzite phase formed. At elevated temperatures the decomposition of the monolithic Ti-Al-N coatings and of the Ti-Al-N / Al-Ti-(Ru)-N multilayer coatings was retarded when they were deposited at low bias voltages ($U_B = -20$ V, -40 V). The accelerated decomposition of the Ti-Al-N based coatings that were deposited at high bias voltages ($U_B = -60$ V, -80 V, -120 V) resulted in a larger volume fraction of wurtzite phase as compared to the coatings deposited at low bias voltages ($U_B = -20$ V, -40 V). This resulted in a significant hardness reduction.

According to the hardness measurements and microstructure analysis of the as-deposited and annealed coatings based on Ti-Al-N, the following application areas are proposed: Ti-Al-N coatings with an average [Al] / Σ[Me] ratio of ~ 0.47 and Ti-Al-N / Al-Ti-(Ru)-N multilayer coatings with a medium [Al] / Σ[Me] ratio of ~ 0.53 that were deposited at low bias voltages (e.g. U_B = -40 V) can be recommended for the application at elevated temperatures up to 950°C, since the formation of large volume fractions of wurtzite phase is retarded and the hardness is maintained after annealing. For applications below the decomposition temperature (< 850°C), Ti-Al-N / Al-Ti-(Ru)-N multilayer coatings with a medium [Al] / Σ[Me] ratio of ~ 0.53 and Ti-Al-N coatings with an average [Al] / Σ[Me] ratio up to 0.56 that were deposited at high bias voltages (e.g. U_B = -80 V) could be favoured. These coatings would provide a high hardness in the as-deposited state and after annealing below < 850°C. Above 850°C a significant hardness loss is expected due to an accelerated decomposition forming high volume fractions of w-AlN.

The microstructure modification via a different ratio of film forming ions to neutrals that impact on the growing film was investigated on the example of Cr-Al-Si-N coatings which were deposited by UBM and HIPIMS with different HIPIMS power applied to the Cr target. During HIPIMS deposition an increased fraction of ionized film forming species is provided as compared to the UBM deposition. Due to that, a higher mobility of the adatoms is achieved during HIPIMS deposition which results in increased crystallite and cluster sizes. An increased ion fraction contributes to the formation of lattice defects that contribute to an increase of the macroscopic lattice deformation and a high hardness of Cr-Al-Si-N coatings deposited by HIPIMS.

As a result of a low ion radiation during UBM deposition, a lower defect density is generated in the Cr-Al-Si-N coatings yielding a low compressive stress and low indentation hardness. On the other hand, the presence of a lower density of point defects admitted the antiferromagnetic ordering of the magnetic moments of the Cr atoms at room temperature which was deduced from the expansion of the fcc elementary cell of (Cr,Al)N and the inverse anisotropy of the elastic lattice deformation.

Finally, it can be stated that the energetic treatment of the growing film, which can be modified by the variation of the bias voltage or by the variation of the ion to neutral flux ratio of film forming species, is a very important approach in order to alter the microstructure and properties of transition metal based nitride films. This could be shown for Ti-Al-N based coatings and Cr-Al-Si-N coatings.

Appendix

A.1 Preferred orientation

Simulated pole figures

The theoretical pole figures that were simulated for a perfect $\{511\}_{fcc}$ and $\{311\}_{fcc}$ in-plane and out-of plane texture for fcc materials are given in Fig. A. 1.

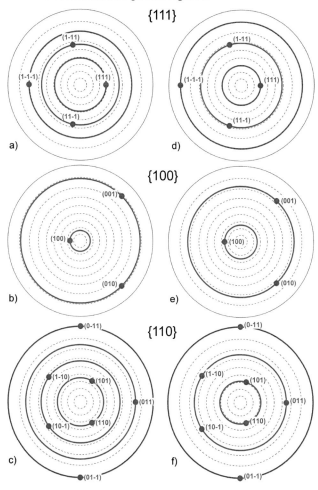

Fig. A. 1: Theoretical pole figures {111}, {100} and {110} that were simulated for a {511} (a,b,c) and {311} (d,e,f) texture in fcc materials. The dashed grid lines correspond to the inclination angle ψ with an interval of 10°. The red symbols denote the poles in case of an in-plane texture and the red lines correspond to the positions of the increased intensity in case of an out-of-plane texture.

Ion channeling directions

The number of atom columns per area are visualized for fcc-TiN in Fig. A. 2 for three different directions.

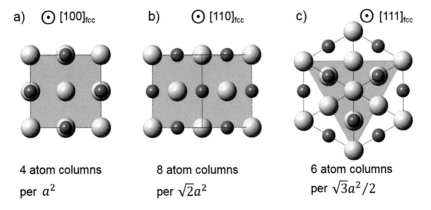

<table>
<tr><td>a) ⊙ [100]_{fcc}</td></tr>
</table>

a) ⊙ $[100]_{fcc}$ b) ⊙ $[110]_{fcc}$ c) ⊙ $[111]_{fcc}$

| 4 atom columns | 8 atom columns | 6 atom columns |
| per a^2 | per $\sqrt{2}a^2$ | per $\sqrt{3}a^2/2$ |

Fig. A. 2: Fcc-TiN plotted in the $[100]_{fcc}$, $[110]_{fcc}$ and $[111]_{fcc}$ direction visualizing the number of atom columns per area (red marked). The relative numbers of atom columns per area a^2 for the $[100]_{fcc}$, $[110]_{fcc}$ and $[111]_{fcc}$ direction correspond to $1:\sqrt{2}:\sqrt{3}$. The size of the red marked area is given at the bottom of the figure. The blue lines indicate the unit cell. The size of the atoms are not to scale.

Surface energy and strain energy

According to Pelleg et al. [217] the surface energy can be expressed as Eq. (A.1) on the basis of the sublimation energy (6.5×10^{-19} J/atom):

$$E^{hkl}_{surface} = 6.5 \times 10^{-19} n_{hkl} /z \; \text{Jcm}^{-2} \qquad (A.1)$$

where n_{hkl} is the number of broken TiN bonds per cm^2 for different orientations and z is the coordination number which equals 6 in case of TiN.

The strain energy is calculated for the **two-dimensional** case according to Ref. [217] as:

$$E^{hkl}_{strain} = \varepsilon^2 E(1-\nu) \qquad (A.2)$$

where ε corresponds to the strain in the film and E corresponds to the mean elastic moduli acting in the plane. The strain energy E^{hkl}_{strain} increases linearly with the film thickness [217].

Pole figure of an fcc-TiN coating deposited by CAE at $U_B = -80$ V

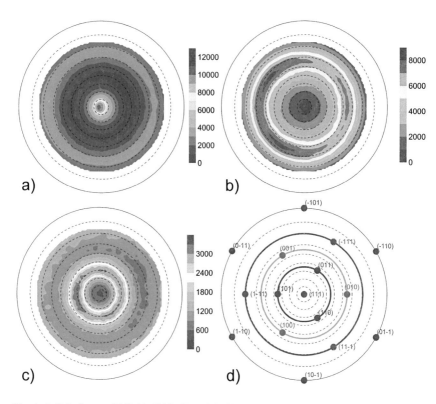

Fig. A. 3: Pole figures {111} (a), {100} (b) and {110} (c) of the fcc-TiN phase measured in a TiN monolithic coating deposited at $U_B = -80$ V using CAE. The dashed grid lines correspond to the inclination angle ψ with an interval of 10°. Simulated stereographic projection for a {111} oriented fcc crystal (d). The coloured circles correspond to the positions of the increased intensity in case of an out-of-plane texture.

A.2 Thermal stability of Ti-Al-N / Al-Ti-(Ru)-N coatings

Parts of the fitted GAXRD patterns of the Ti-Al-N / Al-Ti-Ru-N multilayers of coating series III which were deposited at $U_B = -40$ V and $U_B = -80$ V and annealed at 850°C and 950°C are given in Fig. A. 4.

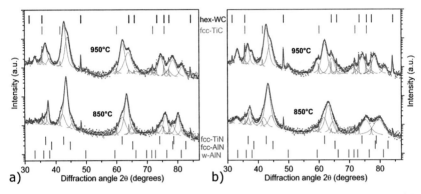

Fig. A. 4: Low angle parts of the GAXRD patterns of the Ti-Al-N / Al-Ti-Ru-N multilayers of the coatings series III deposited at $U_B = -40$ V (a) and $U_B = -80$ V after annealing at 850°C and 950°C. The measured data are shown as blue points and the whole fit as red lines. The grey lines show the individual diffraction peaks as fitted by Pearson VII functions. The positions of the fcc-TiN, fcc-AlN and w-AlN diffraction lines are indicated at the bottom of the figures. The positions of fcc-TiC and hex-WC coming from the substrate are shown at the top.

The thermally activated microstructure changes in the Ti-Al-N / Al-Ti-(Ru)-N multilayers deposited at $U_B = -20$ V and $U_B = -60$ V are summarized in Fig. A. 5.

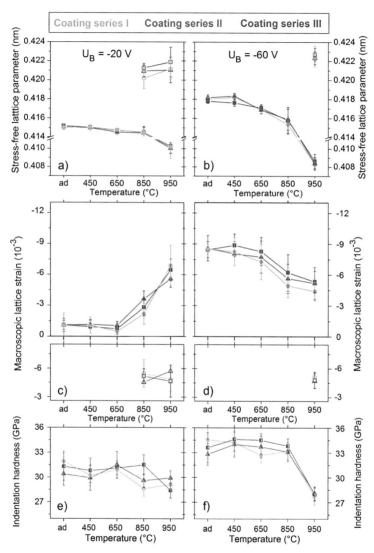

Fig. A. 5: Microstructure evolution of the Ti-Al-N / Al-Ti-(Ru)-N multilayers of the three coating series deposited at $U_B = -20$ V (left column) and $U_B = -60$ V (right column) in the course of the thermal treatment showing the stress-free lattice parameter (a,d) and macroscopic lattice strain (b,e) of the fcc-phases as well as the indentation hardness (c,f). The filled and open symbols represent the microstructure parameters of the fcc-(Ti,Al)N and nearly Al-free fcc-(Ti,Al)N, respectively.

A.3 Heteroepitaxial growth of w-AlN on $(1\bar{1}0)_{fcc}$ and $(00\bar{1})_{fcc}$ oriented fcc-TiN

RSMs of symmetrical reflections

The RSMs showing the $(2\bar{2}0)_{fcc}$ and $(00\bar{2})_{fcc}$ reciprocal lattice points of the fcc-MgO substrate and the fcc-TiN seed layer which were measured in the TiN / AlN / TiN layer stacks deposited onto the $(1\bar{1}0)_{fcc}$ and $(00\bar{1})_{fcc}$ oriented MgO substrates, respectively, are given in Fig. A. 6.

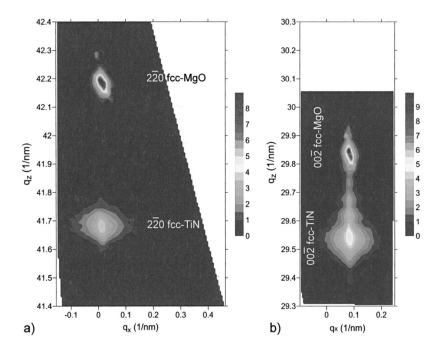

Fig. A. 6: RSMs of the $(2\bar{2}0)_{fcc}$ (a) and $(00\bar{2})_{fcc}$ (b) reciprocal lattice points obtained from the fcc-TiN seed layer and the fcc-MgO substrate which were measured in the TiN / AlN / TiN layer stack deposited onto the $(1\bar{1}0)_{fcc}$ and $(00\bar{1})_{fcc}$ oriented fcc-MgO substrate using HRXRD, respectively. The colour scale refers to the intensity given as natural logarithm. The sample orientation for the RSMs was as follows: $\vec{q}_z \parallel [1\bar{1}0]_{MgO}$ and $\vec{q}_x \parallel [00\bar{1}]_{MgO}$ in figure (a) and $\vec{q}_z \parallel [00\bar{1}]_{MgO}$ and $\vec{q}_x \parallel [1\bar{1}0]_{MgO}$ in figure (b).

Both RSMs were obtained by HRXRD. The $(2\bar{2}0)_{fcc}$ and $(00\bar{2})_{fcc}$ reciprocal lattice points of the fcc-TiN seed layer are slighty broadened as compared to the MgO substrate. The FWHM of the rocking curve measured for the $(2\bar{2}0)_{fcc}$ reciprocal lattice point in the TiN layer deposited onto the $(1\bar{1}0)_{fcc}$ oriented fcc-MgO substrate corresponded to 0.048 °. The FWHM of the rocking curve measured for the $(00\bar{2})_{fcc}$ reciprocal lattice point in the TiN layer deposited onto the

$(00\overline{1})_{fcc}$ oriented fcc-MgO substrate corresponded to 0.054 °. If one assumes that just the lattice planes of the fcc-TiN seed layer contribute to the intensity in the RSMs and in the rocking curves, the FWHM of the rocking curves would indicate a low dislocation density in the range of 10^8 cm^{-2} in both of the fcc-TiN seed layers.

Rocking curve of w-AlN

Due to the low intensity, the measurement of the RSM of the w-AlN layer with the HR diffractometer was not possible. The XRD rocking curve measured at $2\theta_B = 66.06°$ of the 103 diffraction line of w-AlN, which was measured on the HR diffractometer from Seifert FPM without using any detector slit, is shown in Fig. A. 7. The rocking curve of the 013 diffraction line, which was measured with the sample position shown in Fig. 7.48c, revealed the existence of two diffraction maxima in the rocking curve which are separated by $\Delta\omega \approx 4°$. It confirmed the rocking curve measured on the diffractometer D8 ADVANCE.

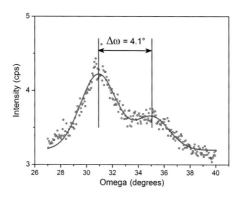

Fig. A. 7: XRD rocking curve of the TiN / AlN / TiN layer stack measured at $2\theta_B = 66.06°$ for $\vec{q} \perp$ $[110]_{MgO}$. The blue symbols represent the measured curve and the red line is the fit of two observed maxima. The maxima are separated by approx. 4.1°. The sample position during the measurement of the XRD rocking curve corresponds to Fig. 7.48c.

RSMs of asymmetrical reflections

The RSMs showing the $(3\overline{3}\overline{1})_{fcc}$ and $(3\overline{3}\overline{1})_{fcc}$ reciprocal lattice points of the fcc-MgO substrate and the fcc-TiN seed layer which were measured in the TiN / AlN / TiN layer stacks deposited onto the $(1\overline{1}0)_{fcc}$ and $(00\overline{1})_{fcc}$ oriented MgO substrates, respectively, are given in Fig. A. 8.

If it is assumed that just the lattice planes of the fcc-TiN seed layer contribute to the reciprocal lattice point of fcc-TiN shown in Fig. A. 8, the q_x position of the reciprocal lattice points of fcc-TiN with respect to the reciprocal lattice point of fcc-MgO indicates that the fcc-TiN seed layer is strained in both TiN / AlN / TiN layer stacks.

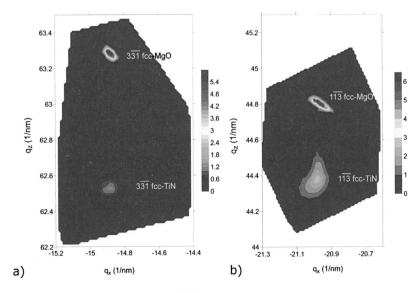

a)

b)

Fig. A. 8: RSMs of the $(3\bar{3}\bar{1})_{fcc}$ (a) and $(1\bar{1}\bar{3})_{fcc}$ (b) reciprocal lattice points obtained from the fcc-TiN seed layer and the fcc-MgO substrate which were measured in the TiN / AlN / TiN layer stack deposited onto the $(1\bar{1}0)_{fcc}$ and $(00\bar{1})_{fcc}$ oriented fcc-MgO substrate using HRXRD, respectively. The colour scale refers to the intensity given as natural logarithm. The sample orientation for the RSMs was as follows: $\vec{q}_z \parallel [1\bar{1}0]_{MgO}$ and $\vec{q}_x \parallel [00\bar{1}]_{MgO}$ in figure (a) and $\vec{q}_z \parallel [00\bar{1}]_{MgO}$ and $\vec{q}_x \parallel [1\bar{1}0]_{MgO}$ in figure (b).

RSM of the MgO substrate

The RSM of the $(2\bar{2}0)_{fcc}$ reciprocal lattice point of the fcc-MgO substrate, which was measured on the D8 Advanced diffractometer from Bruker is shown in Fig. A. 9. Since the ratio of the Cu $K\alpha_2$ to Cu $K\alpha_1$ radiation is 0.15, both contributions are visible in the RSM (see white arrows in Fig. A. 9). Additionally, another area with increased intensity is apparent in the RSM. The origin of this artefact is not clear. This artefact was also observed in the RSMs shown in Fig. 7.48b and Fig. 7.55b, which were obtained from the MgO substrate coated with the TiN / AlN / TiN layer stack. The RSM of the uncoated MgO substrate (see Fig. A. 9) demonstrates that this artefact arises independently from the TiN / AlN / TiN film.

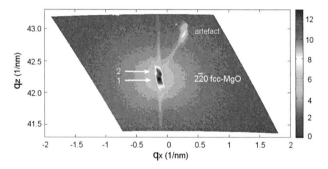

Fig. A. 9: Reciprocal space map of an uncoated $(1\bar{1}0)$ oriented fcc-MgO single crystal measured with the D8 Advanced diffractometer from Bruker. The $(2\bar{2}0)$ reciprocal lattice point and an artefact of unknown origin are visible. The arrows indicate the contribution of the Cu $K\alpha_1$ radiation (arrow 1) and the Cu $K\alpha_2$ radiation (arrow 2). The sample was oriented as follows: $\vec{q_z} \parallel [1\bar{1}0]_{MgO}$ and $\vec{q_x} \parallel [00\bar{1}]_{MgO}$. The colour scale refers to the intensity given as natural logarithm.

Simulated electron diffraction patterns containing the OR $(\overline{1}10)_{fcc}$ ‖ $(\overline{2}10)_w$ and $[001]_{fcc}$ ‖ $[001]_w$

The simulated electron diffraction patterns which were observed in the samples FIB_2 and FIB_1 are shown in Fig. A. 10. The cyan spots mark the position of the diffraction spots which would be expected for the OR characterized by $(\overline{1}10)_{fcc}$ ‖ $(\overline{2}10)_w$ and $[001]_{fcc}$ ‖ $[001]_w$ (see also Eq. (7.19)). However, the SAED measured in the AlN layer of the samples FIB_2 (see Fig. 7.46a) and FIB_1 (see Fig. 7.52a) did not indicate the presence of this OR described by $(\overline{1}10)_{fcc}$ ‖ $(\overline{2}10)_w$ and $[001]_{fcc}$ ‖ $[001]_w$.

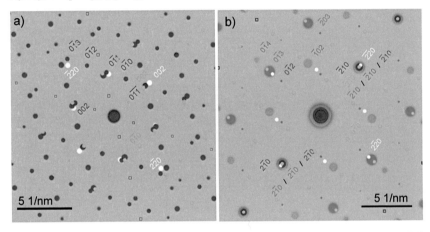

Fig. A. 10: Simulated theoretical electron diffraction pattern of the w-AlN layer of FIB sample 2 (a) and FIB sample 1 (b). The reduced electron diffraction pattern of the fcc-TiN layer is shown as white spots. The coloured spots in Fig. (a) correspond to the electron beam directions $[\overline{1}00]_w$ (red, cyan, blue) and $[\overline{1}\overline{1}0]_{fcc}$ (white). The coloured spots in Fig. (b) correspond to the electron beam directions $[241]_w$ (orange), $[362]_w$ (green), $[121]_w$ grey, $[00\overline{1}]_w$ (cyan) and $[00\overline{1}]_{fcc}$ (white). The cyan coloured diffraction spots are attributed to the OR characterized by $[001]_{fcc}$ ‖ $[001]_w$ and $(1\overline{1}0)_{fcc}$ ‖ $(\overline{2}10)_w$.

Model of the w-AlN layer grown on $(00\bar{1})$ oriented fcc-TiN seed layer

In order to relate the model given in Fig. 7.70 to the crystallographic directions in w-AlN, a sketch showing the alignment of the w-AlN crystals with the orientations "w4.1", "w4.2", "w4.3", "w4.4", "w4.5" and "w4.6" with respect to the $(00\bar{1})$ oriented fcc-TiN seed layer is given in Fig. A. 11. The $[0\bar{1}0]_{fcc}$ direction of the fcc-TiN seed layer points out of the plane of the paper and corresponds to the zone axis used for the TEM investigation (SAED, HRTEM) of the sample FIB_4.

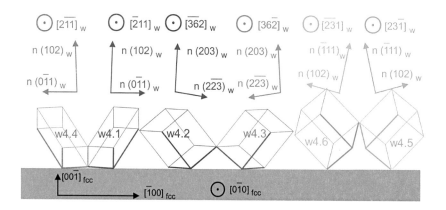

Fig. A. 11: Sketch of the w-AlN elementary cells showing the alignment of the orientation types "w4.1", "w4.2", "w4.3", "w4.4", "w4.5" and "w4.6" with respect to the $(00\bar{1})$ oriented fcc-TiN seed layer as seen in the sample FIB_4. The a, b, c axes of the w-AlN elementary cells are plotted bold.

Bibliography

[1]   S. PalDey, S.C. Deevi, *Single layer and multilayer wear resistant coatings of (Ti,Al)N: a review*, Materials Science and Engineering: A, 342 (2003) 58-79.

[2]   S. Veprek, M.J.G. Veprek-Heijman, *Industrial applications of superhard nanocomposite coatings*, Surface and Coatings Technology, 202 (2008) 5063-5073.

[3]   E. Lugscheider, K. Bobzin, S. Bärwulf, T. Hornig, *Oxidation characteristics and surface energy of chromium-based hardcoatings for use in semisolid forming tools*, Surface and Coatings Technology, 133-134 (2000) 540-547.

[4]   E. Spain, J.C. Avelar-Batista, M. Letch, J. Housden, B. Lerga, *Characterisation and applications of Cr-Al-N coatings*, Surface and Coatings Technology, 200 (2005) 1507-1513.

[5]   O. Knotek, M. Böhmer, T. Leyendecker, *On the structure and properties of sputtered Ti and Al based hard compound films*, Journal of Vacuum Science & Technology A, 4 (1986) 2695-2701.

[6]   H.A. Jehn, S. Hofmann, V.-E. Rückborn, W.-D. Münz, *Morphology and properties of sputtered (Ti,Al)N layers on high speed steel substrates as a function of deposition temperature and sputtering atmosphere*, Journal of Vacuum Science & Technology A, 4 (1986) 2701-2705.

[7]   W.D. Münz, *Titanium aluminium nitride films: a new alternative to TiN coatings*, Journal of Vacuum Science & Technology A, 4 (1986) 2717-2725.

[8]   S. Hofmann, H.A. Jehn, *Oxidation behavior of CrN_x and $(Cr,Al)N_x$ hard coatings*, Werkstoffe und Korrosion, 41 (1990) 756-760.

[9]   O. Knotek, F. Löffler, H.J. Scholl, *Properties of arc-evaporated CrN and (Cr, Al)N coatings*, Surface and Coatings Technology, 45 (1991) 53-58.

[10]   D. Rafaja, M. Dopita, M. Růžička, V. Klemm, D. Heger, G. Schreiber, M. Šíma, *Microstructure development in Cr-Al-Si-N nanocomposites deposited by cathodic arc evaporation*, Surface and Coatings Technology, 201 (2006) 2835-2843.

[11]   D.M. Mattox, *Handbook of Physical Vapor Deposition*, Elsevier, Burlington, 2010.

[12]   A.P. Ehiasarian, R. New, W.D. Münz, L. Hultman, U. Helmersson, V. Kouznetsov, *Influence of high power densities on the composition of pulsed magnetron plasmas*, Vacuum, 65 (2002) 147-154.

[13]   A.P. Ehiasarian, Y.A. Gonzalvo, T.D. Whitmore, *Time-Resolved Ionisation Studies of the High Power Impulse Magnetron Discharge in Mixed Argon and Nitrogen Atmosphere*, Plasma Processes and Polymers, 4 (2007) S309-S313.

[14]   P.J. Kelly, R.D. Arnell, *Magnetron sputtering: a review of recent developments and applications*, Vacuum, 56 (2000) 159-172.

[15]   B. Window, N. Savvides, *Charged particle fluxes from planar magnetron sputtering sources*, Journal of Vacuum Science & Technology A, 4 (1986) 196-202.

[16]   V. Kouznetsov, K. Macák, J.M. Schneider, U. Helmersson, I. Petrov, *A novel pulsed magnetron sputter technique utilizing very high target power densities*, Surface and Coatings Technology, 122 (1999) 290-293.

[17]   U. Helmersson, M. Lattemann, J. Bohlmark, A.P. Ehiasarian, J.T. Gudmundsson, *Ionized physical vapor deposition (IPVD): A review of technology and applications*, Thin Solid Films, 513 (2006) 1-24.

[18]   A. Anders, *A review comparing cathodic arcs and high power impulse magnetron sputtering (HiPIMS)*, Surface and Coatings Technology, 257 (2014) 308-325.

[19]   A. Anders, *Cathodic Arcs: From Fractal Spots to Energetic Condensation*, Springer, New York, 2008.

[20]   A. Anders, *Unfiltered and Filtered Cathodic Arc Deposition*, in: P.M. Martin (Ed.) *Handbook of Deposition Technologies for Films and Coatings: Science, Applications and Technology*, Burlington, 2009, pp. 466-531.

[21]   D.M. Sanders, A. Anders, *Review of cathodic arc deposition technology at the start of the new millennium*, Surface and Coatings Technology, 133–134 (2000) 78-90.

[22]   L. Hägg, C.O. Reinhold, J. Burgdörfer, *Energy gain of highly charged ions in front of LiF*, Nuclear Instruments and Methods in Physics Research Section B, 125 (1997) 133-137.

[23]   J.R. Peterson, *Partial pressure of TiCl₄ in CVD of TiN*, Journal of Vacuum Science & Technology, 11 (1974) 715-718.

[24]   W.D. Sproul, *Very high rate reactive sputtering of TiN, ZrN and HfN*, Thin Solid Films, 107 (1983) 141-147.

[25]   J.E. Sundgren, H.T.G. Hentzell, *A review of the present state of art in hard coatings grown from the vapor phase*, Journal of Vacuum Science & Technology A, 4 (1986) 2259-2279.

[26]   B. Navinsek, *Improvement of Cutting Tools by TiN PVD Hard Coating*, Materials and Manufacturing Processes, 7 (1992) 363-382.

[27]   J. Vogel, E. Bergmann, *Problems encountered with the introduction of ion plating to large-scale coating of tools*, Journal of Vacuum Science & Technology A, 4 (1986) 2731-2739.

[28]   J.M. Molarius, A.S. Korhonen, E. Harju, R. Lappalainen, *Comparison of cutting performance of ion-plated NbN, ZrN, TiN and (Ti,Al)N coatings*, Surface and Coatings Technology, 33 (1987) 117-132.

[29]   P.C. Johnson, H. Randhawa, *Zirconium nitride films prepared by cathodic arc plasma deposition process*, Surface and Coatings Technology, 33 (1987) 53-62.

[30]   E. Ertürk, H.-J. Heuvel, H.-G. Dederichs, *CrN and (Ti,Al)N coatings deposited by the steered arc and random arc techniques*, Surface and Coatings Technology, 39-40 (1989) 435-444.

[31]   K. Wasa, S. Hayakawa, *Structures and resistive properties of sputtered Ti-Zr-Al-N thin films*, Thin Solid Films, 10 (1972) 367-375.

[32]   O. Knotek, W. Bosch, T. Leyendecker, Presentation at 7th International Conference on Vacuum Metallurgy, Linz, Austria, 1985.

[33]   W.D. Münz, J. Göbel, Presentation at 7th International Conference on Vacuum Metallurgy, Linz, Austria, 1985.

[34]   W. Kalss, A. Reiter, V. Derflinger, C. Gey, J.L. Endrino, *Modern coatings in high performance cutting applications*, International Journal of Refractory Metals and Hard Materials, 24 (2006) 399-404.

[35]   M. Kathrein, C. Michotte, M. Penoy, P. Polcik, C. Mitterer, *Multifunctional multi-component PVD coatings for cutting tools*, Surface and Coatings Technology, 200 (2005) 1867-1871.

[36]   R. Rachbauer, A. Blutmager, D. Holec, P.H. Mayrhofer, *Effect of Hf on structure and age hardening of Ti-Al-N thin films*, Surface and Coatings Technology, 206 (2012) 2667-2672.

[37]   K. Kutschej, N. Fateh, P.H. Mayrhofer, M. Kathrein, P. Polcik, C. Mitterer, *Comparative study of $Ti_{1-x}Al_xN$ coatings alloyed with Hf, Nb, and B*, Surface and Coatings Technology, 200 (2005) 113-117.

[38]   D. Rafaja, A. Poklad, V. Klemm, G. Schreiber, D. Heger, M. Šíma, *Microstructure and hardness of nanocrystalline $Ti_{1-x-y}Al_xSi_yN$ thin films*, Materials Science and Engineering: A, 462 (2007) 279-282.

[39]   H. Riedl, D. Holec, R. Rachbauer, P. Polcik, R. Hollerweger, J. Paulitsch, P.H. Mayrhofer, *Phase stability, mechanical properties and thermal stability of Y alloyed Ti-Al-N coatings*, Surface and Coatings Technology, 235 (2013) 174-180.

[40]   A. Knutsson, M.P. Johansson, L. Karlsson, M. Odén, *Thermally enhanced mechanical properties of arc evaporated $Ti_{0.34}Al_{0.66}N$ / TiN multilayer coatings*, Journal of Applied Physics, 108 (2010) 044312.

[41]   M. Stüber, L. Hultman, A. Matthews, *Special Issue of Surface and Coatings Technology on 25 Years of TiAlN Hard Coatings in Research and Industry*, Surface and Coatings Technology, 257 (2014).

[42]   Q. Chen, B. Sundman, *Thermodynamic assessment of the Ti-Al-N system*, Journal of Phase Equilibria, 19 (1998) 146-160.

[43]   Y. Setsuhara, T. Suzuki, Y. Makino, S. Miyake, T. Sakata, H. Mori, *Phase variation and properties of (Ti, Al)N films prepared by ion beam assisted deposition*, Surface and Coatings Technology, 97 (1997) 254-258.

[44]   A. Hörling, L. Hultman, M. Odén, J. Sjölén, L. Karlsson, *Thermal Stability of arc evaporated high aluminium-content $Ti_{1-x}Al_xN$ thin films*, Journal of Vacuum Science & Technology A, 20 (2002) 1815-1823.

[45]   T. Ikeda, H. Satoh, *Phase formation and characterization of hard coatings in the Ti-Al-N system prepared by the cathodic arc ion plating method*, Thin Solid Films, 195 (1991) 99-110.

[46]   U. Wahlström, L. Hultman, J.E. Sundgren, F. Adibi, I. Petrov, J.E. Greene, *Crystal growth and microstructure of polycrystalline $Ti_{1-x}Al_xN$ alloy films deposited by ultra-high-vacuum dual-target magnetron sputtering*, Thin Solid Films, 235 (1993) 62-70.

[47]   M. Zhou, Y. Makino, M. Nose, K. Nogi, *Phase transition and properties of Ti-Al-N thin films prepared by r.f.-plasma assisted magnetron sputtering*, Thin Solid Films, 339 (1999) 203-208.

[48]   D. Rafaja, C. Wüstefeld, M. Dopita, M. Motylenko, C. Baehtz, C. Michotte, M. Kathrein, *Crystallography of phase transitions in metastable titanium aluminium nitride nanocomposites*, Surface and Coatings Technology, 257 (2014) 26-37.

[49]   B.E. Warren, X-ray Diffraction, Dover, New York, 1990.

[50]   C. Schimpf, M. Motylenko, D. Rafaja, *Quantitative description of microstructure defects in hexagonal boron nitrides using X-ray diffraction analysis*, Materials Characterization, 86 (2013) 190-199.

[51]   P.H. Mayrhofer, D. Music, J.M. Schneider, *Influence of the Al distribution on the structure, elastic properties, and phase stability of supersaturated $Ti_{1-x}Al_xN$*, Journal of Applied Physics, 100 (2006) 094906.

[52]   R.F. Zhang, S. Veprek, *Metastable phases and spinodal decomposition in $Ti_{1-x}Al_xN$ system studied by ab initio and thermodynamic modeling, a comparison with the TiN-Si_3N_4 system*, Materials Science and Engineering A, 448 (2007) 111-119.

[53]   A. Kimura, H. Hasegawa, K. Yamada, T. Suzuki, *Effects of Al content on hardness, lattice parameter and microstructure of $Ti_{1-x}Al_xN$ films*, Surface and Coatings Technology, 120-121 (1999) 438-441.

[54]   E.A. Santana, A. Karimi, V.H. Derflinger, A. Schütze, *The role of hcp-AlN on hardness behavior of $Ti_{1-x}Al_xN$ nanocomposite during annealing*, Thin Solid Films, 469-470 (2004) 339-344.

[55]   D. Rafaja, M. Šíma, V. Klemm, G. Schreiber, D. Heger, L. Havela, R. Kužel, *X-ray diffraction on nanocrystalline $Ti_{1-x}Al_xN$ thin films*, Journal of Alloys and Compounds, 378 (2004) 107-111.

[56]   A. Hörling, L. Hultman, M. Odén, J. Sjölén, L. Karlsson, *Mechanical properties and machining performance of $Ti_{1-x}Al_xN$-coated cutting tools*, Surface and Coatings Technology, 191 (2005) 384-392.

[57]   D. Rafaja, A. Poklad, V. Klemm, G. Schreiber, D. Heger, M. Šíma, M. Dopita, *Some consequences of the partial crystallographic coherence between nanocrystalline domains in Ti-Al-N and Ti-Al-Si-N coatings*, Thin Solid Films, 514 (2006) 240-249.

[58]   D. Rafaja, C. Wüstefeld, C. Motylenko, C. Schimpf, T. Barsukova, M.R. Schwarz, E. Kroke, *Interference phenomena in (super)hard nitride nanocomposites: from coatings to bulk materials*, Chemical Society Reviews, 41 (2012) 5081-5101.

[59]   H. Oettel, R. Wiedemann, S. Preißler, *Residual stresses in nitride hard coatings prepared by magnetron sputtering and arc evaporation*, Surface and Coatings Technology, 74-75 (1995) 273-278.

[60]   A.C. Vlasveld, S.G. Harris, E.D. Doyle, D.B. Lewis, W.D. Münz, *Characterisation and performance of partially filtered arc TiAlN coatings*, Surface and Coatings Technology, 149 (2002) 217-224.

[61]   K. Sato, N. Ichimiya, A. Kondo, Y. Tanaka, *Microstructure and mechanical properties of cathodic arc ion-plated (Al,Ti)N coatings*, Surface and Coatings Technology, 163-164 (2003) 135-143.

[62]   M. Ahlgren, H. Blomqvist, *Influence of bias variation on residual stress and texture in TiAlN PVD coatings*, Surface and Coatings Technology, 200 (2005) 157-160.

[63]   J.F. Ziegler, J.P. Biersack, M.D. Ziegler, *SRIM - The Stopping and Range of Ions in Matter*, SRIM-Software, 2008.

[64]   I.G. Brown, *Vacuum arc ion sources*, Review of Scientific Instruments, 65 (1994) 3061-3081.

[65]   C.V. Falub, A. Karimi, M. Ante, W. Kalss, *Interdependence between stress and texture in arc evaporated Ti-Al-N thin films*, Surface and Coatings Technology, 201 (2007) 5891-5898.

[66]   P.H. Mayrhofer, C. Mitterer, L. Hultman, H. Clemens, *Microstructural design of hard coatings*, Progress in Materials Science, 51 (2006) 1032-1114.

[67]   R. Wiedemann, H. Oettel, *Residual Stresses in Coated Materials*, Conference Proceedings 4th International Conference on Residual Stresses, Society for Experimental Mechanics, Baltimore, 1994.

[68]   Y. Lifshitz, S.R. Kasi, J.W. Rabalais, W. Eckstein, *Subplantation model for film growth from hyperthermal species*, Physical Review B, 41 (1990) 10468.

[69]   P. Patsalas, C. Gravalidis, S. Logothetidis, *Surface kinetics and subplantation phenomena affecting the texture, morphology, stress, and growth evolution of titanium nitride films*, Journal of Applied Physics, 96 (2004) 6234-6246.

[70]   F. Adibi, I. Petrov, L. Hultman, U. Wahlström, T. Shimizu, D. McIntyre, J.E. Greene, J.E. Sundgren, *Defect structure and phase transitions in epitaxial metastable cubic $Ti_{0.5}Al_{0.5}N$ alloys grown on MgO(001) by ultra-high-vacuum magnetron sputter deposition*, Journal of Applied Physics, 69 (1991) 6437-6450.

[71]   F. Rovere, D. Music, S. Ershov, M. to Baben, H.-G. Fuss, P.H. Mayrhofer, J.M. Schneider, *Experimental and computational study on the phase stability of Al-containing cubic transition metal nitrides*, Journal of Applied Physics D, 43 (2010) 035302.

[72]   C. Höglund, B. Alling, J. Birch, M. Beckers, P.O.Å. Persson, C. Baehtz, Z. Czigány, J. Jensen, L. Hultman, *Effects of volume mismatch and electronic structure on the decomposition of ScAlN and TiAlN solid solutions*, Physical Review B, 81 (2010) 224101.

[73]   P.H. Mayrhofer, A. Hörling, L. Karlsson, J. Sjölén, T. Larsson, C. Mitterer, L. Hultman, *Self-organized nanostructures in the Ti-Al-N system*, Applied Physics Letters, 83 (2003) 2049-2051.

[74]   P.H. Mayrhofer, L. Hultman, J.M. Schneider, P. Staron, H. Clemens, *Spinodal decomposition of cubic $Ti_{1-x}Al_xN$: Comparison between experiments and modeling*, International Journal of Materials Research, 98 (2007) 1054-1059.

[75]   A. Knutsson, M.P. Johansson, P.O.A. Persson, L. Hultman, M. Odén, *Thermal decomposition products in arc evaporated TiAlN / TiN multilayers*, Applied Physics Letters, 93 (2008) 143110.

[76]   M. Odén, L. Rogström, A. Knutsson, M.R. Terner, P. Hedstrom, J. Almer, J. Ilavsky, *In situ small-angle x-ray scattering study of nanostructure evolution during decomposition of arc evaporated TiAlN coatings*, Applied Physics Letters, 94 (2009) 053114.

[77]   R. Rachbauer, S. Massl, E. Stergar, D. Holec, D. Kiener, J. Keckes, J. Patscheider, M. Stiefel, H. Leitner, P.H. Mayrhofer, *Decomposition pathways in age hardening of Ti-Al-N films*, Journal of Applied Physics, 110 (2011) 023515.

[78]   L. Chen, K.K. Chang, Y. Du, J.R. Li, M.J. Wu, *A comparative research on magnetron sputtering and arc evaporation deposition of Ti-Al-N coatings*, Thin Solid Films, 519 (2011) 3762-3767.

[79]   R. Rachbauer, J.J. Gengler, A.A. Voevodin, K. Resch, P.H. Mayrhofer, *Temperature driven evolution of thermal, electrical, and optical properties of Ti-Al-N coatings*, Acta Materialia, 60 (2012) 2091-2096.

[80]   R. Rachbauer, D. Holec, P.H. Mayrhofer, *Increased thermal stability of Ti-Al-N thin films by Ta alloying*, Surface and Coatings Technology, 211 (2012) 98-103.

[81]   N. Schalk, C. Mitterer, J. Keckes, M. Penoy, C. Michotte, *Influence of residual stresses and grain size on the spinodal decomposition of metastable Ti$_{1-x}$Al$_x$N coatings*, Surface and Coatings Technology, 209 (2012) 190-196.

[82]   D. Rafaja, C. Wüstefeld, C. Baehtz, V. Klemm, M. Dopita, M. Motylenko, C. Michotte, M. Kathrein, *Effect of internal interfaces on hardness and thermal stability of nanocrystalline Ti$_{0.5}$Al$_{0.5}$N coatings*, Metallurgical and Materials Transactions A, 42 (2011) 559-569.

[83]   C. Wüstefeld, D. Rafaja, M. Dopita, M. Motylenko, C. Baehtz, C. Michotte, M. Kathrein, *Decomposition kinetics in Ti$_{1-x}$Al$_x$N coatings as studied by in-situ X-ray diffraction*, Surface and Coatings Technology, 206 (2011) 1727-1734.

[84]   J.L. Endrino, C. Arhammar, A. Gutíerrez, R. Gago, D. Horwat, L. Soriano, G. Fox-Rabinovich, D. Martín y Marero, J. Guo, J.E. Rubensson, J. Andersson, *Spectral evidence of spinodal decomposition, phase transformation and molecular nitrogen formation in supersaturated TiAlN films upon annealing*, Acta Materialia, 59 (2011) 6287-6296.

[85]   L. Rogström, J. Ullbrand, J. Almer, L. Hultman, B. Jansson, M. Odén, *Strain evolution during spinodal decomposition of TiAlN thin films*, Thin Solid Films, 520 (2012) 5542-5549.

[86]   L.J.S. Johnson, M. Thuvander, K. Stiller, M. Odén, L. Hultman, *Spinodal decomposition of Ti$_{0.33}$Al$_{0.67}$N thin films studied by atom probe tomography*, Thin Solid Films, 520 (2012) 4362-4368.

[87]   A. Knutsson, J. Ullbrand, L. Rogström, N. Norrby, L.J.S. Johnson, L. Hultman, J. Almer, M.P. Johansson Jöesaar, B. Jansson, M. Odén, *Microstructure evolution during the isostructural decomposition of TiAlN - A combined in-situ small angle x-ray scattering and phase field study*, Journal of Applied Physics, 113 (2013) 2131518.

[88]   N. Norrby, M.P. Johansson, R. M'Saoubi, M. Odén, *Pressure and temperature effects on the decomposition of arc evaporated Ti$_{0.6}$Al$_{0.4}$N coatings in continuous turning*, Surface and Coatings Technology, 209 (2012) 203-207.

[89]   N. Norrby, H. Lind, G. Parakhonskiy, M.P. Johansson, F. Tasnádi, L.S. Dubrovinsky, N. Dubrovinskaia, I.A. Abrikosov, M. Odén, *High pressure and high temperature stabilization of cubic AlN in Ti$_{0.60}$Al$_{0.40}$N*, Journal of Applied Physics, 113 (2013) 053515.

[90]   K. Kutschej, P.H. Mayrhofer, M. Kathrein, P. Polcik, R. Tessadri, C. Mitterer, *Structure, mechanical and tribological properties of sputtered Ti$_{1-x}$Al$_x$N coatings with $0.5 \leq x \leq 0.75$*, Surface and Coatings Technology, 200 (2005) 2358-2365.

[91]   J. Musil, H. Hrubý, *Superhard nanocomposite Ti$_{1-x}$Al$_x$N films prepared by magnetron sputtering*, Thin Solid Films, 365 (2000) 104-109.

[92]   D. Rafaja, C. Wüstefeld, M. Dopita, V. Klemm, D. Heger, G. Schreiber, M. Šíma, *Formation of defect structures in hard nanocomposites*, Surface and Coatings Technology, 203 (2008) 572-578.

[93]   C. Wüstefeld, D. Rafaja, V. Klemm, C. Michotte, M. Kathrein, *Effect of the aluminium content and the bias voltage on the microstructure formation in Ti$_{1-x}$Al$_x$N protective coatings grown by cathodic arc evaporation*, Surface and Coatings Technology, 205 (2010) 1345-1349.

[94]   M.P. Johansson Joesaar, N. Norrby, J. Ullbrand, R. M'Saoubi, M. Odén, *Anisotropy effects on microstructure and properties in decomposed arc evaporated Ti$_{1-x}$Al$_x$N coatings during metal cutting*, Surface and Coatings Technology, 235 (2013) 181-185.

[95]   L.A. Donohue, I.J. Smith, W.D. Münz, I. Petrov, J.E. Greene, *Microstructure and oxidation-resistance of Ti$_{1-x-y-z}$Al$_x$Cr$_y$Y$_z$N layers grown by combined steered-arc / unbalanced-magnetron-sputter deposition*, Surface and Coatings Technology, 94-95 (1997) 226-231.

[96]   L.A. Donohue, D.B. Lewis, W.D. Münz, M.M. Stack, S.B. Lyon, H.W. Wang, D. Rafaja, *Influence of low concentrations of chromium and yttrium on the oxidation behaviour, residual stress and corrosion performance of TiAlN hard coatings on steel substrates*, Vacuum, 55 (1999) 109-114.

[97]   P. Holubar, M. Jilek, M. Šíma, *Nanocomposite nc-TiAlSiN and nc-TiN-BN coatings: their applications on substrates made of cemented carbide and results of cutting tests*, Surface and Coatings Technology, 120-121 (1999) 184-188.

[98]   M. Pfeiler, G.A. Fontalvo, J. Wagner, K. Kutschej, M. Penoy, C. Michotte, C. Mitterer, M. Kathrein, *Arc evaporation of Ti-Al-Ta-N coatings: The effect of bias voltage and Ta on high-temperature tribological properties*, Tribology Letters, 30 (2008) 91-97.

[99]   K. Kutschej, P.H. Mayrhofer, M. Kathrein, P. Polcik, C. Mitterer, *Influence of oxide phase formation on the tribological behaviour of Ti-Al-V-N coatings*, Surface and Coatings Technology, 200 (2005) 1731-1737.

[100]   L.A. Donohue, J. Cawley, J.S. Brooks, W.D. Münz, *Deposition and characterization of TiAlZrN films produced by a combined steered arc and unbalanced magnetron sputtering technique*, Surface and Coatings Technology, 74-75 (1995) 123-134.

[101]   M. Kathrein, Patent PCT/AT2008/000231, 2008.

[102]   M. Pohler, C. Michotte, M. Penoy, C. Mitterer, M. Kathrein, *Ruthenium Containing Ti-Al-N Hard Coatings*, Conference Proceedings 17th Plansee Seminar, Reutte, 2009, pp. HM 72/1 - HM72/10.

[103]   M.G. Moreno-Armenta, J. Diaz, A. Martinez-Ruiz, G. Soto, *Synthesis of cubic ruthenium nitride by reactive pulsed laser ablation*, Journal of Physics and Chemistry of Solids, 68 (2007) 1989-1994.

[104]   J. Lin, B. Mishra, J.J. Moore, W.D. Sproul, *A study of the oxidation behavior of CrN and CrAlN thin films in air using DSC and TGA analyses*, Surface and Coatings Technology, 202 (2008) 3272-3283.

[105]   J. Vetter, E. Lugscheider, S.S. Guerreiro, *(Cr:Al)N coatings deposited by the cathodic vacuum arc evaporation*, Surface and Coatings Technology, 98 (1998) 1233-1239.

[106]   M. Kawate, A. Kimura Hashimoto, T. Suzuki, *Oxidation resistance of Cr$_{1-x}$Al$_x$N and Ti$_{1-x}$Al$_x$N films*, Surface and Coatings Technology, 165 (2003) 163-167.

[107]   C. Brecher, G. Spachtholz, K. Bobzin, E. Lugscheider, O. Knotek, M. Maes, *Superelastic (Cr,Al)N coatings for high end spindle bearings*, Surface and Coatings Technology, 200 (2005) 1738-1744.

[108]   I.-W. Park, D.S. Kang, J.J. Moore, S.C. Kwon, J.J. Rha, K.H. Kim, *Microstructures, mechanical properties, and tribological behaviors of Cr-Al-N, Cr-Si-N, and Cr-Al-Si-N coatings by a hybrid coating system*, Surface and Coatings Technology, 201 (2007) 5223-5227.

[109]   D. Rafaja, C. Wüstefeld, M. Dopita, M. Růžička, V. Klemm, G. Schreiber, D. Heger,
         M. Šíma, *Internal structure of clusters of partially coherent nanocrystallites in
         Cr-Al-N and Cr-Al-Si-N coatings*, Surface and Coatings Technology, 201 (2007)
         9476-9484.

[110]   T. Polcar, A. Cavaleiro, *High-temperature tribological properties of CrAlN, CrAlSiN
         and AlCrSiN coatings*, Surface and Coatings Technology, 206 (2011) 1244-1251.

[111]   H.-W. Chen, Y.-C. Chan, J.-W. Lee, J.-G. Duh, *Oxidation behavior of Si-doped
         nanocomposite CrAlSiN coatings*, Surface and Coatings Technology, 205 (2010)
         1189-1194.

[112]   K. Bobzin, N. Bagcivan, P. Immich, S. Bolz, R. Cremer, T. Leyendecker, *Mechanical
         properties and oxidation behaviour of (Al,Cr)N and (Al,Cr,Si)N coatings for cutting
         tools deposited by HPPMS*, Thin Solid Films, 517 (2008) 1251-1256.

[113]   A. Sugishima, H. Kajioka, Y. Makino, *Phase transition of pseudobinary Cr-Al-N films
         deposited by magnetron sputtering method*, Surface and Coatings Technology, 97
         (1997) 590-594.

[114]   Y. Makino, K. Nogi, *Synthesis of pseudobinary Cr-Al-N films with B1 structure by rf-
         assisted magnetron sputtering method*, Surface and Coatings Technology, 98 (1998)
         1008-1012.

[115]   H. Hasegawa, M. Kawate, T. Suzuki, *Effects of Al contents on microstructures of
         $Cr_{1-x}Al_xN$ and $Zr_{1-x}Al_xN$ films synthesized by cathodic arc method*, Surface and
         Coatings Technology, 200 (2005) 2409-2413.

[116]   R. Sanjinés, O. Banakh, C. Rojas, P.E. Schmid, F. Lévy, *Electronic properties of
         $Cr_{1-x}Al_xN$ thin films deposited by reactive magnetron sputtering*, Thin Solid Films,
         420-421 (2002) 312-317.

[117]   T. Weirather, C. Czettl, P. Polcik, M. Kathrein, C. Mitterer, *Industrial-scale sputter
         deposition of $Cr_{1-x}Al_xN$ coatings with $0.21 \leq x \leq 0.74$ from segmented targets*, Surface
         and Coatings Technology, 232 (2013) 303-310.

[118]   Y. Makino, *Prediction of phase change in pseudobinary transition metal aluminum
         nitrides by band parameters method*, Surface and Coatings Technology, 193 (2005)
         185-191.

[119]   R.F. Zhang, S. Veprek, *Phase stabilities and spinodal decomposition in the Cr1-xAlxN
         system studied by ab initio LDA and thermodynamic modeling: Comparison with the
         $Ti_{1-x}Al_xN$ and TiN / Si_3N_4 systems*, Acta Materialia, 55 (2007) 4615-4624.

[120]   A.E. Reiter, V.H. Derflinger, B. Hanselmann, T. Bachmann, B. Sartory, *Investigation
         of the properties of $Al_{1-x}Cr_xN$ coatings prepared by cathodic arc evaporation*, Surface
         and Coatings Technology, 200 (2005) 2114-2122.

[121]   P.H. Mayrhofer, D. Music, T. Reeswinkel, H.G. Fuß, J.M. Schneider, *Structure, elastic
         properties and phase stability of $Cr_{1-x}Al_xN$*, Acta Materialia, 56 (2008) 2469-2475.

[122]   A. Kimura, M. Kawate, H. Hasegawa, T. Suzuki, *Anisotropic lattice expansion and
         shrinkage of hexagonal TiAlN and CrAlN films*, Surface and Coatings Technology,
         169-170 (2003) 367-370.

[123]   E. Martinez, R. Sanjinés, O. Banakh, F. Lévy, *Electrical, optical and mechanical
         properties of sputtered CrN_y and $Cr_{1-x}Si_xN_{1.02}$ thin films*, Thin Solid Films, 447-448
         (2004) 332-336.

[124]   L. Castaldi, D. Kurapov, A. Reiter, V. Shklover, P. Schwaller, J. Patscheider, *High temperature phase changes and oxidation behavior of Cr–Si–N coatings*, Surface and Coatings Technology, 202 (2007) 781-785.

[125]   C.S. Sandu, R. Sanjinés, M. Benkahoul, F. Medjani, F. Lévy, *Formation of composite ternary nitride thin films by magnetron sputtering co-deposition*, Surface and Coatings Technology, 201 (2006) 4083-4089.

[126]   J. Patscheider, T. Zehnder, M. Diserens, *Structure–performance relations in nanocomposite coatings*, Surface and Coatings Technology, 146–147 (2001) 201-208.

[127]   S. Veprek, S. Reiprich, *A concept for the design of novel superhard coatings*, Thin Solid Films, 268 (1995) 64-71.

[128]   J.L. Endrino, S. Palacín, M.H. Aguirre, A. Gutiérrez, F. Schäfers, *Determination of the local environment of silicon and the microstructure of quaternary CrAl(Si)N films*, Acta Materialia, 55 (2007) 2129-2135.

[129]   E.O. Hall, *The deformation and ageing of mild steel: II Characteristics of the Lüders deformation*, Proceedings of the Physical Society. Section B, 64 (1951) 747-753.

[130]   N.J. Petch, *The cleavage strength of polycrystals*, Journal of the Iron and Steel Institute, 174 (1953) 25-28.

[131]   S. Veprek, *The search for novel, superhard materials*, Journal of Vacuum Science & Technology A, 17 (1999) 2401-2420.

[132]   L.M. Corliss, N. Elliott, J.M. Hastings, *Antiferromagnetic Structure of CrN*, Physical Review, 117 (1960) 929-935.

[133]   A. Filippetti, W.E. Pickett, B.M. Klein, *Competition between magnetic and structural transitions in CrN*, Physical Review B, 59 (1999) 7043.

[134]   J.D. Browne, P.R. Liddell, R. Street, T. Mills, *An investigation of the antiferromagnetic transition of CrN*, Physica Status Solidi, 1 (1970) 715-723.

[135]   Y. Tsuchiya, K. Kosuge, Y. Ikeda, T. Shigematsu, S. Yamaguchi, N. Nakayama, *Non-stochiometry and Antiferromagnetic Phase Transition of NaCl-type CrN Thin Films Prepared by Reactive Sputtering*, Materials Transaction, 37 (1996) 121-129.

[136]   K. Suzuki, T. Kaneko, H. Yoshida, Y. Obi, H. Fujimori, *Synthesis of the compound CrN by DC reactive sputtering*, Journal of Alloys and Compounds, 280 (1998) 294-298.

[137]   D. Rafaja, C. Wüstefeld, J. Kutzner, A.P. Ehiasarian, M. Šíma, V. Klemm, D. Heger, J. Kortus, *Magnetic response of (Cr,AlSi)N nanocrystallites on the microstructure of Cr-Al-Si-N nanocomposites*, Zeitschrift für Kristallographie 225 (2010) 599-609.

[138]   M.S. Miao, W.R.L. Lambrecht, *Structure and magnetic properties of MnN, CrN, and VN under volume expansion*, Physical Review B, 71 (2005) 214405.

[139]   A. Filippetti, N.A. Hill, *Magnetic Stress as a Driving Force of Structural Distortions: The Case of CrN*, Physical Review Letters, 85 (2000) 5166.

[140]   B. Alling, T. Marten, I.A. Abrikosov, *Effect of magnetic disorder and strong electron correlations on the thermodynamics of CrN*, Physical Review B, 82 (2010) 184430.

[141]   B. Alling, T. Marten, I.A. Abrikosov, A. Karimi, *Comparison of thermodynamic properties of cubic $Cr_{1-x}Al_xN$ and $Ti_{1-x}Al_xN$ from first-principles calculations*, Journal of Applied Physics, 102 (2007) 044314.

[142]   V. Valvoda, R. Kužel, R. Cerny, D. Rafaja, J. Musil, S. Kadlec, A.J. Perry, *Structural analysis of TiN films by Seemann-Bohlin X-ray diffraction*, Thin Solid Films, 193-194 (1990) 401-408.

[143]   A.J. Perry, V. Valvoda, D. Rafaja, *X-ray residual stress measurement in TiN, ZrN and HfN films using the Seemann-Bohlin method*, Thin Solid Films, 214 (1992) 169-174.

[144]   U. Welzel, J. Ligot, P. Lamparter, A.C. Vermeulen, E.J. Mittemeijer, *Stress analysis of polycrystalline thin films and surface regions by X-ray diffraction*, Journal of Applied Crystallography, 38 (2005) 1-29.

[145]   I.C. Noyan, J.B. Cohen, *Residual Stress Measurement by Diffraction and Interpretation*, Springer Verlag, New York, Berlin, Heidelberg, 1987.

[146]   J.F. Nye, *Physical Properties of Crystals - Their Representation by Tensors and Matrices*, Oxford University Press, Oxford, 1985.

[147]   M. Dopita, C. Wüstefeld, V. Klemm, G. Schreiber, D. Heger, M. Růžička, D. Rafaja, *Residual stress and elastic anisotropy in the Ti-Al-(Si-)N and Cr-Al-(Si-)N nanocomposites deposited by cathodic arc evaporation*, Zeitschrift für Kristallographie Supplemente, 27 (2008) 245-252.

[148]   D. Rafaja, V. Valvoda, R. Kužel, A.J. Perry, J.R. Treglio, *XRD characterization of ion-implanted TiN coatings*, Surface and Coatings Technology, 86-87 (1996) 302-308.

[149]   A.J. Perry, V. Valvoda, D. Rafaja, D.L. Williamson, B.D. Sartwell, *On the residual stress and picostructure of titanium nitride films I: Implantation with argon or krypton*, Surface and Coatings Technology 54/55 (1992) 180-185

[150]   H. Möller, G. Martin, *Elastische Anisotropie und röntgenographische Spannungsmessung*, Mitteilungen aus dem Kaiser-Wilhelm-Institut für Eisenforschung, 21 (1939) 261-269.

[151]   M. Dopita, D. Rafaja, C. Wüstefeld, M. Růžička, V. Klemm, D. Heger, G. Schreiber, M. Šíma, *Interplay of microstructural features in $Cr_{1-x}Al_xN$ and $Cr_{1-x-y}Al_xSi_yN$ nanocomposite coatings deposited by cathodic arc evaporation*, Surface and Coatings Technology, 202 (2008) 3199-3207.

[152]   U. Pietsch, V. Holý, T. Baumbach, *High-Resolution X-Ray Scattering*, Springer-Verlag, New York, 2004.

[153]   D.B. Williams, C.B. Carter, *Transmission Electron Microscopy PART 1 Basics*, Springer Science+Business Media, New York, 2009.

[154]   D.B. Williams, C.B. Carter, *Transmission Electron Microscopy PART 2 Diffraction*, Springer Science+Business Media, New York, 2009.

[155]   D.B. Williams, C.B. Carter, *Transmission Electron Microscopy PART 3 Imaging*, Springer Science+Business Media, New York, 2009.

[156]   D.B. Williams, C.B. Carter, *Transmission Electron Microscopy PART 4 Spectrometry*, Springer Science+Business Media, New York, 2009.

[157]   J. Thomas, T. Gemming, *Analytische Transmissionselektronenmikroskopie - Eine Einführung für den Praktiker*, Springer-Verlag Wien, Wien, 2013.

[158]   *Instructions JEM-2200FS Field Emission Electron Microscope, EM-24630 UHADF*, JEOL Ltd., Tokyo, 2010.

[159]   M. De Graef, M.E. McHenry, *Structure of Materials*, Cambridge University Press, Cambridge, 2007.

[160]   DigitalMicrograph Software, Gatan Inc., Pleasanton, CA, USA.

[161]   Single Crystal Software, CrystalMaker Software Limited, Oxfordshire, UK, 2013.

[162]   P. Stadelmann, Java-EMS: JEMS Software, Lausanne, CH, 2014.

[163]   A.G. Jackson, *Handbook of Crystallography*, Springer-Verlag, New York, 1991.

[164]   *EDS Analysis Help Manual 1.8.2*, Gatan Inc., Pleasanton, California, USA, 2009.

[165]   J.I. Goldstein, *Principles of thin film X-ray microanalysis*, in Introduction to Analytical Electron Microscopy, J.J. Hren, J.I. Goldstein, D.C. Joy (Eds.), Plenum Press, New York, 1979, p. 101.

[166]   M. Motylenko, R. Dzhafarov, C. Wüstefeld, D. Rafaja, *EELS and HRTEM based analysis of interfaces in the system Ti-Al-N*, in preparation.

[167]   R. Dzhafarov, *Charakterisierung der Nahordnungsphänomene anhand der Analyse der N-K Nahkantenstruktur in Multilagen- und Gradientenschichten auf der Basis von (Ti,Al)N*, Diploma thesis, Institute of Materials Science, TU Bergakademie, Freiberg, 2015.

[168]   C.C. Ahn, O.L. Krivanek, *EELS ATLAS*, Gatan, Warrendale, 1983.

[169]   V. Klemm, R. Adam, L. Berger, *Determination of the convergence semi-angle and collection semi-angle of a JEOL JEM 2200FS*, Report, Institute of Materials Science, TU Bergakademie Freiberg, Freiberg, 2011.

[170]   C.C. Ahn, *Transmission Electron Energy Loss Spectroscopy in Materials Science and the EELS Atlas*, WILEY-VCH Verlag GmbH & Co. KGaA Weinheim, Weinheim, 2004.

[171]   R.F. Egerton, *Electron Energy Loss Spectroscopy in the Electron Microscope*, Springer, New York, 2011.

[172]   M.R. Schwarz, M. Antlauf, S. Schmerler, K. Keller, T. Schlothauer, J. Kortus, G. Heide, E. Kroke, *Formation and properties of rocksalt-type AlN and implications for high pressure phase relations in the system Si-Al-O-N*, High Pressure Research, 3 (2013) 22-38.

[173]   T. Mizoguchi, I. Tanaka, M. Kunisu, M. Yoshiya, H. Adachi, W.Y. Ching, *Theoretical prediction of ELNES/XANES and chemical bondings of AlN polytypes*, Micron, 34 (2003) 249-254.

[174]   J.C. Le Bossé, M. Sennour, C. Esnouf, H. Chermette, *Simulation of electron energy loss near-edge structure at the Al and N K edges and Al $L_{2,3}$ edge in cubic aluminium nitride*, Ultramicroscopy, 99 (2004) 49-64.

[175]   M. Sennour, C. Esnouf, *Contribution of advanced microscopy techniques to nanoprecipitates characterization: case of AlN precipitation in low-carbon steel*, Acta Materialia, 51 (2003) 943-957.

[176]   D. Holec, R. Rachbauer, D. Kiener, P.D. Cherns, P.M.F.J. Costa, C. McAleese, P.H. Mayrhofer, C.J. Humphreys, *Towards predictive modeling of near-edge structures in electron energy-loss spectra of AlN-based ternary alloys*, Physical Review B, 83 (2011) 165122.

[177]   G. Radtke, T. Epicier, P. Bayle-Guillemaud, J.C. Le Bossé, *N-K ELNES study of anisotropy effects in hexagonal AlN*, Journal of Microscopy, 210 (2003) 60-65.

[178]   D. Holec, P.M.F.J. Costa, P.D. Cherns, C.J. Humphreys, *Electron energy loss near edge structure (ELNES) spectra of AlN and AlGaN: A theoretical study using the Wien2k and Telnes programs*, Micron, 39 (2008) 690-697.

[179]   *EELS Base Help Manual 1.8.2*, Gatan Inc., Pleasanton, California, USA, 2009.

[180]   M. Taylor, R. Smith, F. Dossing, R. Franich, *Robust calculation of effective atomic numbers: The Auto-Zeff software*, Medical Physics, 39 (2012) 1769-1778.

[181]   DIN EN 1071-2, *Verfahren zur Prüfung keramischer Schichten, Teil 2: Bestimmung der Schichtdicke mit dem Kalottenschleifverfahren*, Beuth Verlag GmbH, Berlin, 2002.

[182]   W.C. Oliver, G.M. *Pharr, An improved technique for determing hardness and elastic modulus using load and displacement sensing indentation experiments*, Journal of Materials Research, 7 (1992) 1564-1580.

[183]   DIN EN ISO 14577-1, *Instrumentierte Eindringprüfung zur Bestimmung der Härte und anderer Werkstoffparameter*, Beuth Verlag GmbH, Berlin, 2002.

[184]   T. Chudoba, N.M. Jennett, *Higher accuracy analysis of instrumented indentation data obtained with pointed indenters*, Journal of Applied Physics D, 41 (2008) 215407.

[185]   O. Durand-Drouhin, A.E. Santana, A. Karimi, V.H. Derflinger, A. Schütze, *Mechanical properties and failure modes of TiAl(Si)N single and multilayer thin films*, Surface and Coatings Technology, 163-164 (2003) 260-266.

[186]   G. Korb, *Process for manufacture of a target for cathodic sputtering*, US4752335, 1988.

[187]   DIN ISO 1832, *Wendeschneidplatten für Zerspanwerkszeuge - Bezeichnung*, Beuth Verlag GmbH, Berlin, 2004.

[188]   S. Nagakura, T. Kusunoki, F. Kakimoto, Y. Hirotsu, *Lattice parameter of the non-stoichiometric compound TiN$_x$*, Journal of Applied Crystallography, 8 (1975) 65-66.

[189]   M.T. Baben, L. Raumann, D. Music, J.M. Schneider, *Origin of the nitrogen over- and understoichiometry in Ti$_{0.5}$Al$_{0.5}$N thin films*, Journal of Physics Condensed Matter, 24 (2012) 155401.

[190]   H. Schulz, K.H. Thiemann, *Crystal structure refinement of AlN and GaN*, Solid State Communications, 23 (1977) 815-819.

[191]   D. Rafaja, C. Wüstefeld, D. Heger, M. Šíma, M. Jílek, *Microstructure of Cathodic Arc Evaporated (Al,Ti)N hard Coatings Deposited at Different Orientations to the Target*, Conference Proceedings 18th International Plansee Seminar, Reutte, 2013, pp. HM49/1-HM49/10.

[192]   M. Pfeiler, K. Kutschej, M. Penoy, C. Michotte, C. Mitterer, M. Kathrein, *The influence of bias voltage on structure and mechanical/tribological properties of arc evaporated Ti-Al-V-N coatings*, Surface and Coatings Technology, 202 (2007) 1050-1054.

[193]   M. Pfeiler, C. Scheu, H. Hutter, J. Schnoller, C. Michotte, C. Mitterer, M. Kathrein, *On the effect of Ta on improved oxidation resistance of Ti-Al-Ta-N coatings*, Journal of Vacuum Science & Technology A, 27 (2009) 554-560.

[194]   D. Waasmaier, A. Kirfel, *New Analytical Scattering-Factor Functions for Free Atoms and Ions*, Acta Crystallographica A51 (1995) 416-431.

[195]   R. Rachbauer, E. Stergar, S. Massl, M. Moser, P.H. Mayrhofer, *Three-dimensional atom probe investigations of Ti-Al-N thin films*, Scripta Materialia, 61 (2009) 725-728.

[196]   A.J. Perry, *A Contribution to the Study of Poisson's ratios and Elastic Constants of TiN, ZrN and HfN*, Thin Solid Films, 193-194 (1990) 463-471.

[197]   O. Knotek, R. Elsing, G. Krämer, F. Jungblut, *On the origin of compressive stress in PVD coatings - an explicative model*, Surface and Coatings Technology, 46 (1991) 265-274.

[198]   H. Ljungcrantz, L. Hultman, J.E. Sundgren, L. Karlsson, *Ion induced stress generation in arc-evaporated TiN films*, Journal of Applied Physics, 78 (1995) 832-837.

[199]   D. Holec, F. Rovere, P.H. Mayrhofer, P.B. Barna, *Pressure-dependent stability of cubic and wurtzite phases within the TiN-AlN and CrN-AlN systems*, Scripta Materialia, 62 (2010) 349-352.

[200]   D. Rafaja, V. Klemm, G. Schreiber, M. Knapp, R. Kužel, *Interference phenomena observed by X-ray diffraction in nanocrystalline thin films*, Journal of Applied Crystallography, 37 (2004) 613-620.

[201]   M. Dopita, *Microstructure and properties of nanocrystalline hard coatings and thin film nanocomposites*, Doctoral Thesis, Faculty of Mathematics and Physics, Charles University Prague, Prague, 2009.

[202]   K. Murgaeva, *TEM-Analyse von Orientierungsbeziehungen und Grenzflächen in Nanokompositen*, Diploma thesis, Institute of Materials Science, TU Bergakademie, Freiberg, TU Bergakademie Freiberg, Freiberg, 2011.

[203]   C. Wüstefeld, M. Motylenko, D. Rafaja, *Crystallography of internal interfaces in TiAlN / AlTi(Ru)N coatings*, 65th Research Conference Freiberg, Freiberger Forschungsheft B359, Freiberg, 2014, pp. 26-30.

[204]   P.H. Mayrhofer, D. Music, J.M. Schneider, *Ab initio calculated binodal and spinodal of cubic $Ti_{1-x}Al_xN$*, Applied Physics Letters, 88 (2006) 071922.

[205]   B. Alling, A.V. Ruban, A. Karimi, O.E. Peil, S.I. Simak, L. Hultman, I.A. Abrikosov, *Mixing and decomposition thermodynamics of $c-Ti_{1-x}Al_xN$ from first-principles calculations*, Physical Review B, 75 (2007) 045123.

[206]   D. Rafaja, V. Valvoda, A.J. Perry, J.R. Treglio, *Depth profile of residual stress in metal-ion implanted TiN coatings*, Surface and Coatings Technology, 92 (1997) 135-141.

[207]   R.R. Manory, A.J. Perry, D. Rafaja, R. Nowak, *Some effects of ion beam treatments on titanium nitride coatings of commercial quality*, Surface and Coatings Technology, 114 (1999) 137-142.

[208]   J.W. Cahn, *On spinodal decomposition*, Acta Metallurgica, 9 (1961) 795-801.

[209]   P.H. Mayrhofer, F.D. Fischer, H.J. Böhm, C. Mitterer, J.M. Schneider, *Energetic balance and kinetics for the decomposition of supersaturated $Ti_{1-x}Al_xN$*, Acta Materialia, 55 (2007) 1441-1446.

[210]   G.K. Williamson, A.H. Hall, *X-Ray Line Broadening from Filed Aluminium and Wolfram*, Acta Metallurgica, 1 (1953) 22-31.

[211]   H. Holleck, V. Schier, *Multilayer PVD coatings for wear protection*, Surface and Coatings Technology, 76–77, Part 1 (1995) 328-336.

[212]   P.E. Hovsepian, D.B. Lewis, W.D. Münz, *Recent progress in large scale manufacturing of multilayer/superlattice hard coatings*, Surface and Coatings Technology, 133-134 (2000) 166-175.

[213]   C. Wüstefeld, M. Motylenko, D. Rafaja, D. Heger, C. Michotte, C. Czettl, M. Kathrein, *Microstructure of TiAlN / AlTiRuN multilayers grown by cathodic arc evaporation*, Conference Proceedings 18th International Plansee Seminar, Reutte, 2013, pp. HM140/1-HM140/13.

[214]   C. Wüstefeld, M. Motylenko, D. Rafaja, C. Michotte, C. Czettl, *Defect engineering in Ti-Al-N based coatings via energetic particle bombardment during cathodic arc evaporation*, in: D. Rafaja (Ed.), *Functional structure design of new high-performance materials via atomic design and defect engineering – ADDE*, SAXONIA Standortentwicklungs- und –verwaltungsgesellschaft mbH, Freiberg , 2015, pp. 200-223.

[215]   F. Tasnádi, I.A. Abrikosov, L. Rogström, J. Almer, M.P. Johansson, M. Odén, *Significant elastic anisotropy in $Ti_{1-x}Al_xN$ alloys*, Applied Physics Letters, 97 (2010) 231902.

[216]   C.H. Ma, J.H. Huang, H. Chen, *Texture evolution of transition-metal nitride thin films by ion beam assisted deposition*, Thin Solid Films, 446 (2004) 184-193.

[217]   J. Pelleg, L.Z. Zevin, S. Lungo, N. Croitoru, *Reactive-sputter-deposited TiN films on glass substrates*, Thin Solid Films, 197 (1991) 117-128.

[218]   I. Petrov, L. Hultman, J.E. Sundgren, J.E. Greene, *Polycrystalline TiN films deposited by reactive bias magnetron sputtering: Effects of ion bombardment on resputtering rates, film composition, and microstructure*, Journal of Vacuum Science & Technology A, 10 (1992) 265-272.

[219]   L. Hultman, J.E. Sundgren, J.E. Greene, D.B. Bergstrom, I. Petrov, *High-flux low-energy (=~20 eV) N+2 ion irradiation during TiN deposition by reactive magnetron sputtering: Effects on microstructure and preferred orientation*, Journal of Applied Physics, 78 (1995) 5395-5403.

[220]   D. Rafaja, A. Poklad, G. Schreiber, V. Klemm, D. Heger, M. Šíma, *On the preferred orientation in $Ti_{1-x}Al_xN$ and $Ti_{1-x-y}Al_xSi_yN$ thin films*, Zeitschrift für Metallkunde / International Journal of Materials Research and Advanced Techniques, 96 (2005) 738 - 742.

[221]   L. Hultman, J.E. Sundgren, J.E. Greene, *Formation of polyhedral N_2 bubbles during reactive sputter deposition*, Journal of Applied Physics, 66 (1989) 536-544.

[222]   M. Marlo, V. Milman, *Density-functional study of bulk and surface properties of titanium nitride using different exchange-correlation functionals*, Physical Review B, 62 (2000) 2899-2907.

[223]   C. Tholander, B. Alling, F. Tasnadí, J.E. Greene, L. Hultman, *Effect of Al subsitution on Ti, Al and N adatom dynamics on TiN(100), (011) and (111) surfaces*, Surface Science, 630 (2014) 28-40.

[224]   D. Gall, S. Kodambaka, M.A. Wall, I. Petrov, J.E. Greene, *Pathways of atomistic processes on TiN(001) and (111) surfaces during film growth: an ab initio study*, Journal of Applied Physics, 93 (2003) 9086-9094.

[225]   A. Karimi, W. Kalss, *Off-axis texture in nanostructured $Ti_{1-x}Al_xN$ thin films*, Surface and Coatings Technology, 202 (2008) 2241-2246.

[226] D. Rafaja, T. Merkewitz, C. Polzer, V. Klemm, G. Schreiber, P. Polcik, M. Kathrein, *Effect of the targets microstructure on the microstructure of the cathodic arc evaporated $Ti_{0.5}Al_{0.5}N$*, Conference Proceedings 17th International Plansee Seminar, Reutte, 2009, pp. HM38/1-HM38/12.

[227] W.D. Sproul, D.J. Christie, D.C. Carter, *Control of reactive sputtering processes*, Thin Solid Films, 491 (2005) 1-17.

[228] A. Béré, A. Serra, *Atomic structures of twin boundaries in GaN*, Physical Review B, 68 (2003) 033305.

[229] A. Béré, A. Serra, *On the atomic structures, mobility and interactions of extended defects in GaN: dislocations, tilt and twin boundaries*, Philosophical Magazine, 86 (2006) 2159-2192.

[230] S. Horiuchi, I. Toshihiko, K. Asakura, *A new type of twin in an AlN crystal*, Journal of Crystal Growth, 21 (1974) 17-22.

[231] T. Mizoguchi, I. Tanaka, S. Yoshioka, M. Kunisu, T. Yamamoto, W.Y. Ching, *First-principles calculations of ELNES and XANES of selected wide-gap materials: Dependence on crystal structure and orientation*, Physical Review B, 70 (2004) 045103.

[232] R. Gindt, R. Kern, *Étude des macles du nitrure d'aluminium. Interprétation causale*, Bulletin de la Société Francaise de Minéralogie et de Cristallographie, 81 (1958) 266-273.

[233] A. Hecimovic, *High Power Impulse Magnetron Sputtering (HIPIMS) - Fundamental Plasma Studies and Material Synthesis,* Doctoral Thesis, Sheffield Hallam University, Sheffield, 2009.

[234] M. Jilek, T. Cselle, P. Holubar, M. Morstein, M.G.J. Veprek-Heijman, S. Veprek, *Development of novel coating technology by vacuum arc with rotating cathodes for industrial production of nc-$(Al_{1-x}Ti_x)N$ / a-Si_3N_4 superhard nanocomposite coatings for dry, hard machining*, Plasma Chemistry and Plasma Processing, 24 (2004) 493-510.

[235] ICSD Database, Version 2009-2, FIZ Karlsruhe & NIST 2009.

[236] A. Hecimovic, A.P. Ehiasarian, *Time evolution of ion energies in HIPIMS of chromium plasma discharge*, Journal of Physics D: Applied Physics, 42 (2009) 135209.

[237] W. Ensinger, *Low energy ion assist during deposition - An effective tool for controlling thin film microstructure*, Nuclear Instruments and Methods in Physics Research, Section B, 127-128 (1997) 796-808.

[238] L. Hultman, U. Helmersson, S.A. Barnett, J.E. Sundgren, J.E. Greene, *Low-energy ion irradiation during film growth for reducing defect densities in epitaxial TiN(100) films deposited by reactive-magnetron sputtering*, Journal of Applied Physics, 61 (1987) 552-555.